FLYING THE PRIVATE PILOT FLIGHT TEST

■ A MANUAL

RON   FOWLER

Flying THE
Private
Pilot
Flight
Test ■ A MANUAL

IOWA STATE UNIVERSITY PRESS / AMES

THIS BOOK IS FOR **LINDSAY,**

ASHLEY,

CLANCY,

C. B.,

KALEY,

AND

JAMES DALTON

■

RON FOWLER is a flight instructor and teaches aviation courses at Valencia Community College, Orlando, Florida, where he is also director of aviation curriculum. He is the author of *Making Perfect Landings in Light Airplanes* and *Making Perfect Takeoffs in Light Airplanes* (Iowa State University Press), *Flying the Commercial Test* and *Preflight Planning* (Macmillan), *Flying Precision Maneuvers in Light Airplanes* (Delacorte Press), and numerous magazine articles.

Illustrations by **Jan Avis**
Photographs by **John Tate**

© 1994 Iowa State University Press, Ames, Iowa 50014
All rights reserved

Authorization to photocopy items for internal or personal use, or the internal or personal use of specific clients, is granted by Iowa State University Press, provided that the base fee of $.10 per copy is paid directly to the Copyright Clearance Center, 27 Congress Street, Salem, MA 01970. For those organizations that have been granted a photocopy license by CCC, a separate system of payments has been arranged. The fee code for users of the Transactional Reporting Service is 0-8138-2487-7/94 $.10.

♾ Printed on acid-free paper in the United States of America

First edition, 1994
  Second printing, 1995
  Third printing, 1996

          Library of Congress Cataloging-in-Publication Data
Fowler, Ron
    Flying the private pilot flight test: a manual/Ron Fowler.—1st ed.
        p.   cm.
    Includes bibliographical references and index.
    ISBN 0-8138-2487-7
    1. Airplanes—Piloting. 2. Private flying—United States—Examinations—Study guides. I. Title.
    TL710.F649   1994
    629.'5217—dc20   93-44365

Contents

Preface ■ *Adventure need not be an act of derring-do.*

In truth, high adventure is anything that you *wish* to call high adventure while you are doing it. The potential for adventure surrounds you in every waking moment, if you only reach for it; and high adventure is so important to a full life. Flying can be that adventure. Many of you who want to become pilots do so to expand your business capabilities—many of you become pilots just for recreation—and there are those of you for whom flying will be a profession as a commercial pilot. These are all valid reasons. But whatever reasoning has led you to the cockpit, I hope a sheer sense of adventure lies within you. Aloft within the cockpit, your high adventure is so close at hand.

Adventure aloft is not confined to places and events, but also a feeling within yourself—first as you learn the discipline and values of flying and then later as you transfer these elements to your work and play. Make no mistake. Whether your life *now* is either singular or indifferent, you *will* look at life from a different perspective as you progress toward being a certified pilot in the cockpit in full control of your airplane.

Will altitude change a pilot's perspective? Most certainly. And it's easy to understand why. Simply put, your thoughts are free to soar as far as the horizon before you—there are no limits. Look at it this way: Confine a person to a cramped little room, set them to thinking, and their thoughts are confined. But perch that same individual high on a pinnacle, give their thoughts stretching room, and their perspective will expand. It's inevitable. Pilots aloft are at their pinnacle, working at the controls behind the singing engine. Their thoughts are not walled in but have a whole sky in which to grow. Pilots can ricochet an idea off a towering cumulus. Or bounce a supposition against the distant, curving horizon. Or project a thought through the night sky to the farthest star. It is a simple matter of close spaces for close ideas and open spaces for open ideas. Take yourself to a new altitude — a new perspective is waiting there.

I envy you as you set out on this adventure of learning to fly and I am pleased that you allow me, through this book, to travel with you. You will be asked to satisfactorily complete the *Private Pilot Practical Test* — your flight and oral test for pilot certification. This practical test is not easy — it is not *intended* to be easy. (And in truth, I don't think you *want* it to be easy as you earn your place in that small, unique society of pilots.) I want to help you successfully perform the flight test and share in your victory. I wrote this book to help you, the rated pilot who wants refresher training or the student pilot, develop the three basic tools of any pilot: knowledge, skill, and judgement.

Adequate aeronautical *knowledge* is the foundation upon which to build piloting skill and judgement. Without adequate aeronautical knowledge, even the most meticulous skill or keenest judgement has limited potential for success in the air. All too often I have seen the most skillful pilots — who lack adequate knowledge — direct their fine mechanical skill in the totally wrong direction. And too often I have seen the most logically thinking pilots — also lacking knowledge — at a loss without adequate information on which to base their judgement.

Not just aeronautical knowledge, but *adequate* aeronautical knowledge has a direct bearing on the flight test. For the most part, I write about the "nuts and bolts," but there is also theory and information as it pertains to each specific flight-test maneuver. This information is critical to your flight test.

The flight test is administered within two frameworks: inflight performance and oral examination. Typically, much of the oral exam is conducted in the examiner's office before the two of you even get near the airplane. Questions concerning the aerodynamic forces and aircraft limitations as they pertain to flight-test maneuvers are fair game.

Also, remember that flight-test examiners rarely remain silent during the inflight portion of the test. They ask questions. When you make a necessary move during a demonstrated maneuver, for example, the examiner might ask: Why did you need to do that? The answer often lies in basic aerodynamics, and you need to know how to reply. The text addresses both this need and how to best answer the examiner's questions.

The driving force behind writing this text is thoroughness, thoroughness, thoroughness. I strive for this element for one cardinal reason. There is an old saying that amateur flying is a great sport. In truth, it may be a great sport, but it may also be an awful experience, filled with frustration and hazard. Professional flying, on the other hand (especially as a private pilot) is filled with pleasure and safety as the pilot gains confidence and skill. And by "professional" I do not mean a pilot who flies for hire or has accumulated thousands of hours aloft. I simply mean a pilot determined to deliver the very best that plane and pilot, as a team, can give. This professionalism comes only through adequate practice and aeronautical knowledge. I strive to tell you everything I know that is significant and that will help you become a safe pilot. I feel that you are *entitled* to anything I know.

The flight test will test your *skill* through demonstrations of maneuvers needed by a certified pilot-in-command. These skills are demonstrated through a series of specified flight maneuvers, flown by you during the test. We will discuss each maneuver at length.

Each chapter is devoted to a single maneuver or closely related maneuvers. The body of each chapter covers the associated maneuvers in depth, the tools to use, the pitfalls to avoid, and the theory behind the procedures.

You will be asked to demonstrate each maneuver in a specified manner and you will be evaluated against specified tolerances for satisfactory performance. Each chapter concludes with flight-test guidelines, in which you are instructed, step by numbered step, just how to fly each flight-test maneuver. Following this 1-2-3 guide through each maneuver, you are shown the specific tolerances against which the examiner will evaluate your performance. In short, you will know exactly what to expect from your examiner, and you will know just what your examiner expects from you.

We all value good, sound judgement and enjoy meeting and working with people who use good judgement. We all are pleased when we ourselves exercise sound judgement. For pilots, good judgement in the cockpit is critical. But I must warn you, there is a subtle (but significant) difference between good judgement *aground* and good judgement *aloft*. Aground, judgement is often tempered and directed by the potential for compromise and negotiation, which is as it should be. Aloft, there is a subtle shift toward the *pragmatic*. Aloft, the dictates, physics, and elements of nature rule—natural laws that often extend no potential for either compromise or negotiation. A pilot's judgement aloft must find its basis within the basic truth of flying, which simply says, Plane and sky accept no excuses and grant no special considerations. It makes no difference whether we walk our Earth as good or bad, wise or foolish, poor or rich. In the cockpit, our differences remain earthbound. We touch the controls as one—a pilot. If our actions are *right*, the flight will probably succeed. If our actions are *not* right, the flight may likely fail.

Judgement aloft, then, is pragmatic; judgement that *knows* the inflexible elements that play out within the cockpit; judgement that requires action without hesitation. Much of what I have to say during our time together is said in an effort to identify these inviolate elements of flight and how you must cope with them. It is our task to work *with* the elements of nature, *not against* them. And in doing so we receive one of the remarkable gifts of flight — the opportunity for kinship with the elements. Flying is one of the few remaining human efforts that allow each of us to become truly a man or woman of nature.

If all of my instructing has taught me anything, it is this: The average individual learning to fly holds the potential for greatness. Excellence is within your reach. I'm 100% behind you; I understand your flying hopes and concerns. Between us we'll get things sorted out, for I expect you to reach your own potential.

Oh, how I envy you and your new adventure. For myself, it truly seems like only a few logbook pages ago, not decades. And inside I still feel like I'm twenty-five. So, let's begin our work together, toward knowledge, skill, and judgement. And let's also enjoy the pleasure of one another's company.

FLYING THE PRIVATE PILOT FLIGHT TEST

■ A MANUAL

1.

Maneuvering at Critically Low Airspeeds

In the private pilot flight test you are asked to demonstrate your ability to maneuver your plane at critically low airspeeds in two areas of operation: stall recognition and recovery, and sustained flight at the plane's minimum controllable airspeed. Most flight-test examiners call for a stall before they ask to see control at minimum airspeed. There is logic behind this sequence. Pilots must recognize the onset of a stall and be able to effect an immediate remedy if they are to successfully sustain flight at minimum controllable airspeed.

STALLS

By the time you are ready for the flight test, you know what a stall is — that condition whereby the wings no longer provide lift sufficient to support the plane's weight, and the plane quits flying. A stall occurs when there is insufficient airflow passing over an adequate area of wing surface. The two requisites — sufficient airflow and adequate lifting area of wing — can easily erode with *either* insufficient airspeed that does not provide an adequate flow of air or an excessive angle of attack that diminishes the available lifting area of the wings.

3

■ Insufficient Airspeed

The typical wing of a light plane is designed as an airfoil that develops 75% of its lift from the air that flows over the upper surface (Fig. 1-1). Airspeed is one of the factors that controls the volume of that airflow. Fly the wing fast — greater airflow. Fly slow, less flow — diminished lift. The ultimate case of a moving wing being stalled by insufficient airspeed can be found in a taxiing airplane. Obviously an extreme angle of attack (Fig. 1-2) is not the culprit here; the wing is riding level. But at the slow taxiing speed, there just isn't the needed flow of air.

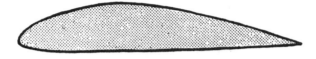

Fig. 1-1.   A typical light-plane airfoil develops most of its lift from the curved upper surface.

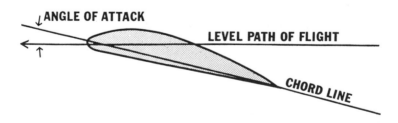

Fig. 1-2.   The angle of attack is simply the angle that exists between the wing's chord line and the airplane's flight path. Here, the plane is maintaining altitude along a level flight path, with the nose slightly elevated.

Once in the air, however, different factors control the airspeed at which the wing stalls. The *density* of the air is one of those factors.

Air Density

Air density is just an expression of how many molecules of air there are in a cubic inch. The fewer the molecules per cubic inch, the faster you must fly to maintain the same number of molecules flowing over the wing; thus, the faster the stall speed.

Three factors affect the air density: altitude, temperature, and moisture.

ALTITUDE. As the altitude increases the atmospheric pressure drops. And, as the pressure drops, there is less pressure to compress the air molecules, so they "spread out" or become less dense. This increases the true airspeed at which the wing will stall by about 2% per 1000 feet. For example, a wing that stalls at a true airspeed (TAS) of 60 knots at sea level, will stall at a TAS of 66 knots at 5000 feet. (Remember that here we are speaking of an increase in *true airspeed*. The *indicated* stall speed remains unchanged regardless of density. This is because the airspeed indicator pitot tube, which senses airflow velocity by air impact, is subject to the same reduced air density as is the wing. Therefore, at any given altitude the airspeed indicator displays a speed about 2% slower per 1000 feet than we are truly flying.

TEMPERATURE. Any increase in temperature increases the true stalling speed in the same manner as an increase in altitude, and for the same reason. Heated air drives the molecules of air farther apart, resulting in less dense air. Imagine a toy balloon filled with air and sealed. Now gently heat the balloon and watch it expand. Obviously, no additional molecules of air slip into the sealed balloon. What *does* happen is the heat increases the distance between the existing molecules. Thus, hot air significantly increases the true stall speed by about 5% for each 30° Fahrenheit (F). Consider this example: the plane stalls at 60 knots TAS at 60°F. Increase the temperature to 90°F and the wing stalls at 63 knots TAS.

MOISTURE. Oddly enough, the moisture content also affects air density and the true airspeed at which the wing stalls. Moisture also increases the distance between the air molecules and thus has a thinning effect. Although not as dramatic as either altitude or temperature, moisture, nevertheless, further aggravates the detrimental effects of the other two. Consider the pilot operating at a 4000-foot mountain-airport elevation on a hot, humid day. Stall can easily occur at a TAS 10 knots greater than expected. Does this mean that the prudent pilot must carry an additional 10 knots on the airspeed indicator when approaching a landing or lifting off? Absolutely not! Remember, thin air affects the airspeed indicator pitot tube exactly the same way it affects the wing's flyability. So, if the handbook stipulates an indicated 80-knot approach speed at sea level, use the same 80 knots indicated on the mountain-airport final approach. Just keep in mind that your *actual* TAS is considerably faster. What does this mean to the pilot? It means that extra speed and momentum at touchdown translate into a much longer descent path and ground roll (Fig. 1-3).

Fig. 1-3. Extra speed and momentum at touchdown translates into a longer descent path and ground roll when landing at a mountain airport.

The same applies on takeoff. If the recommended *indicated* rotate (or flying) speed is 60 knots at sea level, then use that same indication when lifting off at elevation. Again, remember that your actual rotate speed is greater, and it takes more runway to achieve the additional acceleration (Fig. 1-4).

Factors other than altitude, temperature, and density affect stalling speed. Weight of the aircraft is another.

Aircraft Weight

The heavier the plane, the more lift the wings must provide; thus the airspeed must increase to increase the airflow over the wing. The weight of the aircraft (and the stall speed) can be increased by either increasing

4-6

Sport B19

TAKE-OFF DISTANCE — HARD SURFACE

ASSOCIATED CONDITIONS

POWER	FULL THROTTLE
MIXTURE	LEAN TO MAXIMUM RPM, THEN ENRICH SLIGHTLY
FLAPS	UP
RUNWAY	LEVEL, DRY, HARD SURFACE
WEIGHT	2150 LBS

TAKE-OFF SPEEDS

LIFT OFF	70 MPH/61 KTS
50 FT	80 MPH/70 KTS

WIND COMPONENT DOWN RUNWAY KNOTS	SEA LEVEL				2000 FT				4000 FT				6000 FT				8000 FT			
	OAT F	OAT C	GROUND ROLL FEET	TOTAL OVER 50 FT OBSTACLE FEET	OAT F	OAT C	GROUND ROLL FEET	TOTAL OVER 50 FT OBSTACLE FEET	OAT F	OAT C	GROUND ROLL FEET	TOTAL OVER 50 FT OBSTACLE FEET	OAT F	OAT C	GROUND ROLL FEET	TOTAL OVER 50 FT OBSTACLE FEET	OAT F	OAT C	GROUND ROLL FEET	TOTAL OVER 50 FT OBSTACLE FEET
0	23	5	836	1331	16	9	953	1510	9	13	1089	1717	2	17	1247	1956	6	21	1436	2232
	41	5	930	1478	34	1	1062	1680	27	3	1215	1913	20	7	1393	2182	13	11	1606	2495
	59	15	1032	1635	52	11	1178	1861	45	7	1350	2123	38	3	1550	2425	31	9	1784	2777
	77	25	1137	1803	70	21	1303	2055	63	17	1495	2347	56	13	1719	2686	49	9	1981	3078
	95	35	1251	1982	88	31	1435	2262	81	27	1649	✓ 2586	74	23	1899	2963	67	19	2191	3400
15	23	5	642	1195	16	9	738	1361			952	1552	2	17	980	1773	6	21	1132	2030
	41	5	717	1331	34	1	826				953	1733	20	7	1100	1964	13	11	1274	2274
	59	15	799	1476	52	11	921				1063	1928	38	3	1230	2216	31	9	1428	2538
	77	25	886	1631	70	21			63	17	1182	2136	56	13	1370	2452	49	9	1590	2820
	95	35	979	1796	88				81	27	1310	2359	74	23	1520	2711	67	19	1766	3122
30	23	5	471				1237	9	13	638	1415	2	17	743	1671	6	21	867	1860
	41	5	530				617	1383	27	3	720	1584	20	7	840	1817	13	11	982	2089
	59	15				11	693	1539	45	7	808	1765	38	3	945	2028	31	9	1105	2335
	77				70	21	774	1706	63	17	904	1959	56	13	1057	2255	49	9	1239	2600
				1642	88	31	861	1885	81	27	1007	2167	74	23	1180	2498	67	19	1384	2884

Fig. 1-4. Higher true airspeed at rotation requires additional acceleration distance when departing high-elevation airports.

the gross weight with additional passengers, fuel, or baggage; or increasing the aerodynamic load, or gravity (G) forces, imposed any time you maneuver the plane.

INCREASED GROSS WEIGHT. Stall speed increases as the gross weight increases. Typically, the stall speed of a light, single-engine craft increases at about 1 knot for each 100 pounds of useful load. This also means that stall speed *decreases* by about 1 knot per 100 pounds of reduced weight. Thus, a plane with a difference of 500 pounds between a light load and a heavy one may have a difference of 5 knots in the stalling speed—a significant difference. Does this have an impact on piloting? Sure.

For example, let's say you are on final approach to a short runway. The plane is lightly loaded at 500 pounds under the maximum allowable gross weight. Minimizing the landing roll is a concern, and a safe reduction in approach speed will help meet this concern. The aircraft manual states a safe approach speed that is calculated as a multiple of the plane's stall speed with flaps properly set: 1.3 × stall speed = approach speed. This approach speed, however, is commonly calculated for the plane carrying its maximum allowable weight. But the example has the weight at 500 pounds below maximum; therefore, stalling speed is now lower, and you can safely reduce the speed of your single-engine light plane by 1 knot for each 100 pounds below maximum. This reduction in stall speed is significant when approaching a short field, because each 1-knot reduction on final approach means a savings of about 2% in the total distance required to clear a 50-foot obstacle.

INCREASED AERODYNAMIC LOADING. Stall speed is affected by aerodynamic loading, just as it is when the gross weight is increased. Increase the aerodynamic loading (G force) and you increase the speed at which the wings will stall. G force is an expression of *how much* you increase the weight of the plane and its payload through aerodynamic loading, and the G forces are imposed anytime the plane is deflected from its existing flight path. That is, a force of 1 G — as in straight and level flight — imposes no extra loading. A 2-G maneuver, however, has the plane and everything in it weighing twice its normal weight. Leveling from a descent is a maneuver that imposes loading. The steeper the descent and the more rapid the recovery, the greater the G load and the greater the stalling speed. The message here is, Do not make derring-do pull-outs from fast descents. (There is another hazard to be aware of, apart from entering a high-speed stall. Normal category, light aircraft are built to withstand stress to 3.8 Gs; utility aircraft to 4.4 Gs. Exceed the stress limits of your aircraft with an excessively rapid pull-out and you may leave a trail of fluttering aluminum in your wake. This sort of air play is reserved for aerobatic planes that have stress tolerances up to 6 Gs — and pilots who have the training to go along with it.)

Banking through a turn also imposes aerodynamic loading and increased stall speed. The steeper the bank, the greater the G force (Fig. 1-5), and thus, the greater the stall speed (Fig. 1-6). This loading is the result of simple centrifugal force. To understand the concept, relate it to the childhood pastime of looping a bucket of water roundabout and overhead. Centrifugal force prevents the water from pouring out of the upturned bucket. Your arm also tells you that the bucket of water feels much heavier. Figure 1-6 shows just how much stalling speed is increased at various angles of bank. Notice how dramatically stall speed increases as you reach a 40-degree bank. The message is clear: if possible, avoid steep turns within 1000 feet of the ground (where altitude for stall recovery is limited) when flying at slow airspeeds. But what about circumstances in which you *must* turn steeply (perhaps to avoid traffic) while flying low and slow? Do two things to stack the cards in your favor. First, if altitude permits, slightly lower the nose to gain extra flying speed. Second, add power for the turn, for power reduces the stalling speed.

Power vs. Stall Speed

Power affects stall speed. The more power you apply, the more you lower the stalling speed. The difference between power-off and power-on stall speed is significant. In the typical single-engine light plane, for example, the difference is between 7 and 10 knots.

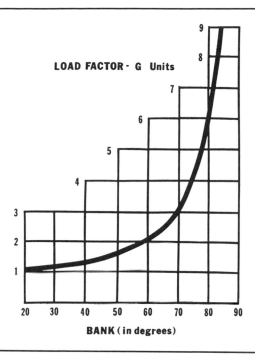

LOAD FACTOR - G Units

BANK (in degrees)

Fig. 1-5. The steeper the bank, the greater the aerodynamic loading (G force, or units).

STALL SPEEDS – MPH CAS

Gross W.	ANGLE OF BANK			
CONDITION	0°	20°	40°	60°
Flaps UP	55	57	63	78
Flaps 20°	49	51	56	70
Flaps 40°	48	49	54	67
POWER OFF — AFT CG				

Fig. 1-6. Due in large part to extra aerodynamic loading, stall speed increases with bank.

The reason for this reduction in stalling speed is quite simple. When under a lot of power, the propeller acts like a giant fan blowing a large mass of fast-moving air over the inboard surface of the wing. In a slow-moving plane, this acts to supplement a low airspeed. This fact is accounted for in the plane's takeoff and landing performance charts. Each manufacturer's recommended *rotate* and *approach* speeds (Fig. 1-7) are calculated to provide a safe margin above stall. Yet the *recommended* rotate speed is usually several knots *lower* than the recommended approach speed. This does not mean that the manufacturer is suggesting a smaller safety margin on takeoff, but that they predicate the rotate speed with the assumption of a full-power lift-off, which results in a lowered stall speed. And, they assume a reduced-power final approach will be made with its higher stall speed.

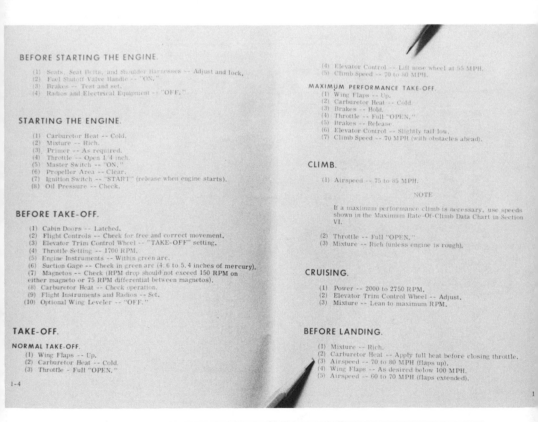

Fig. 1-7. The difference between rotate and approach speeds requires different power applications.

Two lessons come immediately to mind that you can gain from this information. First, you learn why it is so important to lower the nose and add all available power when the plane enters the region of an inadvertent stall; a lowered nose results in extra airspeed and power for windflow. Second, you learn not to chop power on a slow final approach, for you might stall a wing. It is far better to lower the nose, slowly reduce power, and accept a longer landing—or simply go around and try again.

In review, the factors that affect the plane's stalling speed are (1) altitude, (2) temperature, (3) moisture content, (4) gross weight, (5) aerodynamic loading, and (6) power.

Don't hesitate to discuss your understanding of these factors as you and the flight-test examiner talk about stalls. Many applicants feel that the flight test is primarily a showcase of their piloting skills. Nothing could be further from the truth. Much of the flight-test time is given to conversation before flying even starts. The examiner is interested in discovering your depth of aeronautical knowledge. The examiner knows full well the three basic tools that a pilot takes to the cockpit: knowledge, judgement, and skill. And of these three, *aeronautical knowledge* is the cornerstone. For without adequate knowledge, pilots have nothing on which to base their *judgement*. Too often, unknowledgeable pilots direct their otherwise superb *skill* in the wrong direction for the circumstance at hand.

However, examiners are not mind readers. You must tell them what you know. You may even need to sketch some line drawings to supplement your verbal explanations. If so, grab a pencil and pad and draw them. Or take this book to your flight test. The drawings are simple, but illustrate the concepts adequately. They are meant for your use, so feel free to do so whenever they can support your explanations. (Do examiners feel that reference to printed material is cheating? Absolutely not! In fact, they are quite interested in seeing what reference materials you rely on. Examiners know that a good aviation library is essential. With this in mind, there is a list of suggested reading materials at the back of this book. Do not hesitate to show the examiner any of your references. And if you do show your references, don't be surprised if the examiner recommends further titles with which to expand your aviation knowledge.)

Thus far we have discussed one condition that causes a wing to stall—insufficient airspeed, which does not provide adequate airflow. Now let's move to another stall condition—angle of attack, which diminishes the available lifting surface of the wing.

■ Angle of Attack

The angle of attack is the angle at which the plane's flight attitude strikes the mass of air through which it is flying (Fig. 1-2). Within limitations, the more you increase this angle of attack at any given airspeed, the more you increase the curvature of the airflow over the wing's upper surface (Fig. 1-8), which results in increased lift.

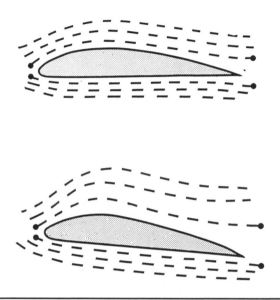

Fig. 1-8. An increase in the angle of attack results in a greater curvature of airflow across the upper surface of the wing.

This extra lifting ability is based on Bernoulli's (Ber nu′ leez) theorem. Bernoulli discovered that faster flowing fluids exert less pressure on the surface over which they flow than do slower moving fluids. How does this theorem apply to a wing's lift? A wing is curved on the top and flat on the bottom. Picture the leading edge of the wing striking and separating two molecules of air; one passes over the wing, the other slips beneath (Fig. 1-9). Both molecules will reach the trailing edge of the wing at the same time. Which molecule travels the greater distance in the same time span? Which molecule travels faster? Obviously, the molecule passing over the wing's curved upper surface travels farther and faster in the same time span than does the molecule following a straight line along the underside of the wing. The faster windflow over the wing's upper

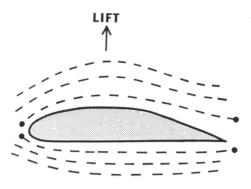

Fig. 1-9. The faster airflow across the upper surface of the wing helps to create lift.

surface creates a lower pressure. The slower wind under the lower surface results in a relatively higher pressure beneath the wing. Nature tries to fill the "low" above from the "high" beneath, but the wing stands in the way. Therefore, the wing must move with the pressure flow as physics tries to gain equilibrium—hence lift.

It is easy to visualize how this added curvature increases the differential between the wing's upper and lower pressure readings and produces added lift at any given airspeed. But, remember that there are *limits* to the amount of angle of attack that you can apply without actually destroying lift. These limits are related to the area of lifting surface and a situation called "critical" angle of attack.

Critical Angle of Attack

Every wing design has its own limit-of-attack angle—a critical angle beyond which lift erodes. This angle has little to do with the airspeed being flown or the attitude in which the airplane is flying. The wing's critical angle remains virtually the same both at high or low speeds; in a dive, turning, or in a climb; and when either heavily or lightly loaded.

The wing's critical angle of attack is met when the airflow is restricted from flowing smoothly over a sufficient area of the upper wing surface. This situation is brought about by a factor called the "separation point."

Separation Point

Remember that about 75% of the wing's lift is derived from airflow across the upper surface; the up-tilted lower surface provides only about

25%, by impact, as it planes its way across the ocean of air passing beneath. (These percentages change somewhat in favor of the lower surface as the angle of attack increases. But you depend primarily on the airflow acting on the wing's upper surface.)

Unfortunately, as you increase the angle of attack, you diminish the upper wing's area of effective lifting surface. As you tilt the wing higher and higher, the point where the airflow breaks contact with the wing's upper surface is ever closer to the leading edge of the wing (Fig. 1-10). This point of "break away" is called the separation point. Increasing the angle of attack excessively moves the separation point too far forward and prevents sufficient lifting area. At this angle, you will have reached

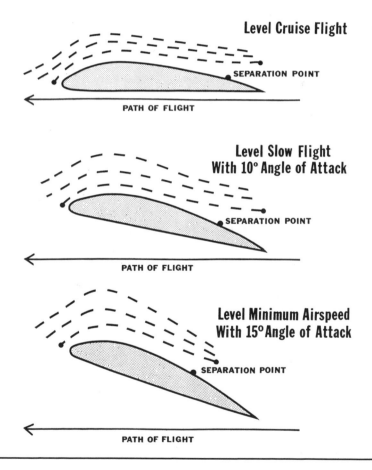

Fig. 1-10. As the angle of attack increases, the separation point moves forward to diminish the lifting area of the wing.

the airfoil's *critical angle of attack* (between 12 and 17 degrees in most light planes), and the wing stalls.

The critical angle of attack is the basis for the old saying, The wing can stall at *any* airspeed, at *any* attitude. To better understand this, imagine yourself in a steep dive at high speed. Can you stall the plane in this nose-down attitude and fast airspeed? Certainly! All you need to do is yank the stick back to deflect the wings 12 to 17 degrees away from the flight path (Fig. 1-11). The separation point jumps toward the leading edge of the wing, and the wing stalls.

Two lessons can be learned from this imaginary high-speed stall. First, don't yank on the controls, for this imposes aerodynamic loading

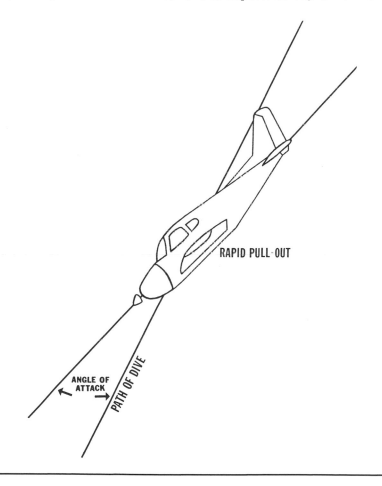

Fig. 1-11. A wing can stall at any airspeed or attitude, even in a high-speed dive, if you misuse the controls.

and displaces the separation point. Second, regardless of your attitude or airspeed, if you detect a potential stall, move the stick steadily forward to put the separation point back where it belongs. (More about detecting a stall later.)

So, you need to worry not only about low airspeed when it comes to stalls, but also about the *separation point* and *critical angle.* There are times, however, when you need to fly slowly with a high angle of attack, such as during the final approach to a landing. And wouldn't it be great if there was a device that allowed this high angle of attack yet kept the separation point in place? Fortunately, you *do* have such a device — flaps.

Flaps

Flaps help to solve an important low-speed dilemma. We are all too familiar with the trade-off between airspeed and angle of attack when maintaining sufficient lift to support the weight of the aircraft. When flying slow, you must increase the angle of attack to increase the curvature in order to replace lift lost to a diminished airflow. In so doing, however, the separation point begins moving forward to reduce the wing's lifting area. That's where flaps come into the picture.

Extended flaps create an aerodynamic increased angle of attack, while allowing the separation point to hold its position well aft of the wing's leading edge. Here is how it works: Picture the plane riding level with flaps extended. Now draw an imaginary line from the trailing edge of the extended flaps upward and forward through the leading edge of the wing (Fig. 1-12). The angle at which that line crosses the flight path is the aerodynamic angle of attack created by the extended flaps. Yet, notice that the separation point atop the level wing has not moved,

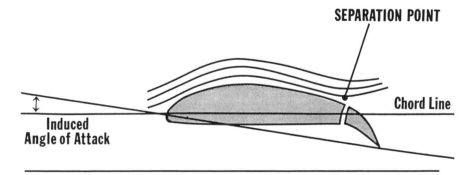

Fig. 1-12. Extended flaps create an aerodynamic angle of attack without displacing the separation point.

simply because the wing rides level. Thus, you have gained the lifting advantage of an increased angle of attack while not losing any lifting surface. This is why most manufacturers recommend that full flaps be used when landing.

Flaps represent one set of controls that affect stall characteristics. Ailerons and the rudder are two more. If you do not coordinate these two controls, you can seriously affect the available lifting surface of the wings.

■ Coordinating the Controls

Control of coordination will be discussed in depth in chapter 6. But for our discussion of stalls, let's, for the moment, accept three statements:

1. When you bank the wings with ailerons, you must coordinate that action with an equal rudder force. (You will see later *how* to assure this coordination.)

2. If you fail to coordinate the aileron and rudder actions, the plane will slide sidewise through the air. (Again, you will see later exactly *why* this happens.)

3. If you allow the plane to slide sidewise through the air, the available lifting surface of the wing is diminished.

How is that lifting area diminished? A wing is designed with a precise curvature on the upper surface. This curvature is designed to allow the close flow of air to contact the wing surface until the air reaches the normally anticipated separation point. When the leading edge of the wing is forced sidewise through the air, however, the wind must cross the upper surface at a slant (Fig. 1-13). As the air slants across the surface, it follows a slightly different wing curvature, causing airflow to begin to ripple and eddy too soon. Contact with the wing is destroyed long before it reaches the normal separation point, thus diminishing the available lifting area.

What significance does this have? Let's say you are flying close to the ground, turning base to final in the landing approach. You let the approach speed get a bit low and then fail to coordinate the rudder with the ailerons during the turn. That plane could stall.

This example demonstrates the prime reason—to prevent an altered stall characteristic—for learning the technique and instinct for control coordination from your very first hour at the controls of an airplane.

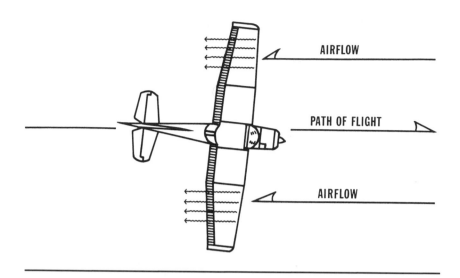

Fig. 1-13. If we force the wings to fly sidewise through the air, the wind must flow across the wings at a slant.

■ Designing Stall Characteristics

Modern light planes are designed so they possess "mellow" stall characteristics. That is, they are designed to give the pilot ample warning that a stall is about to occur. And once stalled, they are designed to produce "mild" symptoms—a stall devoid of violent pitching or rolling. Their design also allows an easy recovery with little loss of altitude. You should be prepared to discuss these design features with your examiner. The design involves wing design and aircraft loading.

Wing Design

Light-plane wings are designed so they do not stall all at once. This design has the wingtips still flying long after the wing roots have stalled, which results in a fair warning to the pilot that a full stall is on the way, modest pitch and roll when the stall breaks, and easy recovery. Design elements such as wingtip washout, placement of ailerons, wing shape, wing slots, and stall strips are employed to bring all this about.

WINGTIP WASHOUT. The wingtips of many light planes are designed so they have a forward and downward slope (Fig. 1-14). This is called wingtip washout. As you can see from Figure 1-14, when the main wing section has reached its critical angle of attack and stalls, the tip has

Fig. 1-14. The wing tips of many light planes are designed with a forward, downward slope.

a few degrees of flying left in it. This design feature provides three advantages to the pilot. First, it allows early warning of an impending stall, for it forces the inner wing section (root) to stall first. When the wing root stalls, the disturbed airflow tumbles rearward to shake the horizontal stabilizer. This produces the "buffeting," which warns the pilot that the stall has started and is working its way out to the wingtip. (Fig. 1-15).

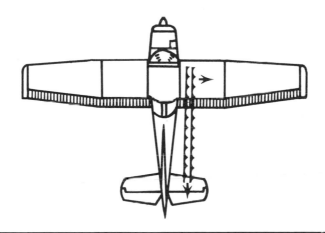

Fig. 1-15. The "buffet" you feel prior to the stall is caused by the disturbed air tumbling across the wing roots to shake the horizontal tailplane.

Second, wingtip washout spreads the stall out; that is, the wing stalls by segments, which prevents a violent maneuver that would occur if the entire wing stalled at once. The wings roll rather slowly, and the nose sinks relatively gently. (Aerobatic planes, on the other hand, are designed without wingtip washout. But, the aerobatic pilot *wants* a fast-breaking stall with which to accomplish some of the snappier airshow stunts.)

Third, wingtip washout facilitates an easier and quicker stall recovery. In most cases the pilot can recover before the entire wing is stalled; a quick recovery to normal flight is assured when recovery action is initiated.

PLACEMENT OF AILERONS. Ailerons on light planes are located near the wingtips — within the wing section that is the last to stall. This greatly facilitates stall recovery, for the pilot maintains roll control with which to counter any wing drop that may occur with the stall. (Modern plane designs are required to demonstrate 10% aileron control at the point of full stall.)

WING SHAPE. Light-plane wings are generally designed as either rectangular or slightly tapered (Fig. 1-16).

Both shapes are intended to keep the wingtip flying after the rest of the wing has given up. Generally, the rectangular design does a better job in this regard, by providing more lifting area at the tip than does the tapered wing. But, while cheaper to build, the rectangular wing also produces more drag. Therefore, the rectangular design is most often encountered on the less-expensive, lower performance light planes.

The tapered wing, while more costly than a rectangular design, develops less drag, yet provides reasonably adequate wingtip area for the ailerons. It is used with the faster light-plane models.

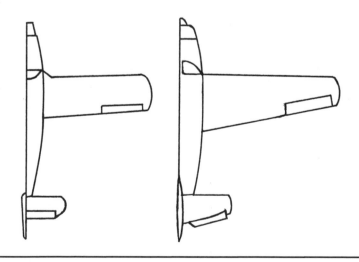

Fig. 1-16. Rectangular and tapered wings generally used on light aircraft provide good lift at the tips.

Light-plane manufacturers stay away from swept-back wings and those that employ elliptical shapes (Fig. 1-17). These designs greatly reduce drag, but they provide minimal wingtip area. Additionally, for structural integrity, the ailerons are usually placed well inboard of the wingtip. For these reasons, the stall characteristics of these designs are less than desirable. Swept-back and elliptical wings are most often encountered on very high performance military or jet aircraft. And believe me, pilots who fly them do *not* go up for an afternoon of stall recovery practice. Quite to the contrary, they passionately avoid any situation that even approaches a stall situation.

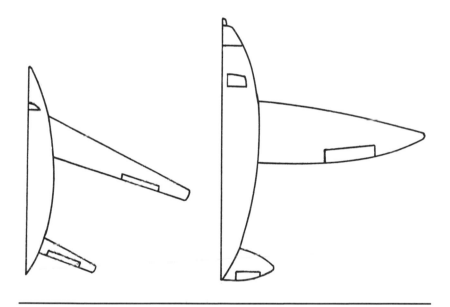

Fig. 1-17. The slender tips of swept-back and elliptical wings provide little lift. To preserve structural strength, ailerons are usually mounted well inboard of the tips.

WING SLOTS. There is another stall-delaying device — wing slots. However, wing slots are expensive to build into the wing. They are designed to re-introduce a second flow of air across the upper surface of the wing at a high angle of attack, to keep the separation point in place longer (Fig. 1-18). You normally find these slots on special-purpose aircraft — those with the need to fly very slowly.

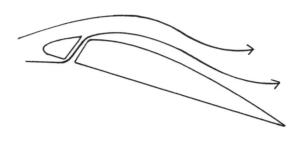

Fig. 1-18. Wing slots allow a second flow of air with which to keep the separation point in place.

STALL STRIPS. There is a very inexpensive way to make the wingtip provide lift longer than the wing root; *force* the root to stall as the wing angles upward. This is easily accomplished with stall strips, which are merely small metal "lift spoilers" attached to the leading edge of the wing root. As you see in Figure 1-19, these strips force an early breakup of the airflow across the root when the angle of attack increases toward the point of stall.

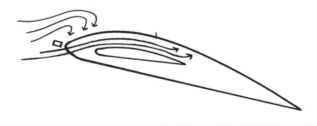

Fig. 1-19. Stall strips force an early break-up of airflow at the wing roots.

You should know how the wing of your plane is designed to produce a docile stall. Don't hesitate to discuss the design features with your flight-test examiner, for this shows your depth of understanding.

■ Stall Awareness and Prevention

Pilots often wonder how a stall can occur accidentally. After all, your stall practice clearly shows the full warnings of an on-coming stall—the sight of a low-airspeed indication or pitched nose, the sound of the stall horn or the fading rush of slipstream across the canopy, the feel of sagging aileron control or the pull of excess G force. You might wonder how an accidental stall can slip up on you unnoticed: The answer in a word—distraction.

Nearly every accidental approach to a stall that I have witnessed was caused by momentary distraction, which made the pilot forget the number one priority—control. Examples are easy to imagine: A pilot turning base to final, still searching out the closing traffic that is on a long straight-in. Or a pilot, while executing a go-round in an out-of-trim plane, getting concerned with communicating with the tower. Or even the pilot flying low and slow, then turning too steeply to inspect a point of interest on the ground. You must try to avoid distractions.

Basic aircraft control simply means that the pilot is in positive control of the plane's three basic elements—altitude, heading, and airspeed. Unless you have these elements under exact control, you have no business attending to other matters such as communicating with a controller, sorting out navigation, or even attending to an aircraft malfunction. For if distraction robs you of basic aircraft control, a stall is a distinct possibility.

Yet, we are only human. And we know that distractions *can* and *do* occur, no matter how vigilant we are. Therefore, to further guard against an unintentional stall at low altitudes (where they hold the greatest hazard), consider putting these ten rules of thumb into play when you fly within 1000 feet of the ground.

1. Avoid flying at speeds lower than the plane's "flaps-up" approach speed. (Figure this minimum speed as $1.3 \times$ flaps-up stall speed.)

2. Avoid any maneuvering that requires a bank in excess of 30 degrees. (To avoid excess aerodynamic loading.)

3. Run your engine at a setting no lower than 55% of power. (To assure greater airflow across the wing.)

4. Keep one hand on throttle, so it is already there if you need it. (Time can be critical in preventing a stall.)

5. Fly with the flaps extended one notch. (To provide a greater cushion between flying speed and stalling speed.)

6. Run the engine with the carburetor heat full hot. (Be sure to lean the fuel mixture to guard against the over-rich condition that carburetor heat produces.)

7. Avoid pitch attitudes that cover the horizon with the cowl. (If you can't *see* the horizon, you can only guess at the pitch attitude.)

8. Don't study a ground object that lies behind your shoulders. (If you need a better look, turn the airplane.)

9. Make certain that the plane is trimmed properly for the airspeed

being flown. (An out-of-trim plane just begs a pilot to let their attention wander.)

10. Cut cockpit chatter to a minimum and avoid dealing with an aircraft discrepancy at low altitudes. (The plane's attitude, heading, and airspeed demand your fullest attentions.)

There is one final thing that you must keep foremost in your mind when reacting to an inadvertent stall at low altitude. You *must* conform to your trained reflex actions rather than fall victim to your natural instinct. A fully developed stall may point the nose toward the ground. Trained reflex demands that you apply forward pressure on the yoke to quickly break the stall. Yet, natural instinct may shout, Pull up! Get it away from the ground! Your throttle hand has been trained to add power to get the plane flying again. Yet, instinct may tell you to slow the action, reduce power. Many good pilots have come to grief simply because panic has let them succumb to natural instinct in a moment of crisis, rather than to put trained reflex into action. Panic in a moment of crisis is your worst enemy. When flying an airplane, *force* yourself to keep thinking right through that moment.

Another facet of reaction to stalls that often inhibits prompt corrective action by the pilot is an illogical anxiety toward the stall attitude. If this is true in your case, you are not alone. Let me recount an incident that occurred a few years ago — it may put your concerns to rest.

■ The Stall Attitude — Mental and Angular

The summer morning breeze puffing through the open hangar brought a whiff of creosote from the heating tarmac ramp. It was still cool in the shaded corner where I sat at my oil-stained wooden desk, but I knew the afternoon students were in for some uncomfortable hours.

Several uncowled airplanes, including a perfectly restored little tailwheel J-3, were parked in different stages of tear-down, and the pleasant ring of a dropped wrench drifted across the hangar. Outside, a mechanic was running up the engine in one of my three trainers. (It was the 4-year-old ship that we referred to as "the new plane.")

The other two planes were out on solo cross-countries, and I'd spent most of the morning trying to figure out how to squeeze an unpalatable fuel increase into a palatable hourly rental increase. I was no closer to a solution when Tom Cutter walked in. He had to wait a moment for the little engine on the other side of the metal hangar wall to rev down before he could speak.

"Say, Ron, you got a minute?" he asked.

"Have a seat," I said, as I scooped the pile of old aviation magazines from the metal folding chair.

Tom Cutter had earned his pilot certificate 3 years earlier at another airport and now had about 200 hours. He enjoys flying on the weekends and recently bought his first plane—a real nice red and white used Cherokee.

Tom sat down, leaned back in the chair, and reached over to toy with the small brass *RON FOWLER—FLYING INSTRUCTOR* desktop sign that for years had been my sole advertising program.

He said, "I've been up giving the Cherokee a workout, and I think I've got a problem with my flying."

"Oh?" I said to fill the pause.

"I think so," he repeated.

"What happened?"

"Well, I was practicing stalls, and I'll tell you, doing those things scares me as much now as when I first learned them."

It's a common anxiety, of course, but Tom didn't know that. He only knew *he* didn't like stalls.

"How can I get over being nervous about them? Or am I just stuck with it forever? I don't even know *why* they worry me."

I think one of the reasons why many pilots often harbor anxiety toward stalls is that they are forced into stall recovery practice before they have had a chance to learn positive aircraft control in normal attitudes. Most students use their common sense in the cockpit. And when they're asked to fly the plane through a stall, they *know* they're dealing with a critical attitude. If they aren't satisfied with their *normal* aircraft control, they know that stall recovery practice is asking for something they are ill prepared to deliver. How much better it would be for the student and instructor alike, if fledglings didn't discover stalls and stall recoveries until *after* they have mastered straight and level, turns, climbs, descents, ground-reference maneuvers, and (except for a few gentle power-off stalls) even landings.

But this was all past history for Tom. He wanted to know how he could cure his anxiety *now*. And the answer was simple. I reached up and took the model Skyhawk from the shelf against the hangar wall.

"Tom," I said, "most pilots don't like stalls because they have a misconception of the stall attitude. Many pilots hold an exaggerated opinion of the plane's pitch when the stall occurs." I handed the model to Tom. "Show me what you think the pitch attitude looks like when the stall breaks."

"About like so," he answered, holding the model's nose pitched up about 50 degrees.

"Nope. You have this nose pitched way too high. It's probably this exaggerated idea of the stall attitude that's causing your anxiety."

"Why is that?"

"Because your *idea* of the stall's pitch attitude leads you to a couple of misconceptions of what the plane *might do* when the stall breaks."

Tom looked at the model airplane and said, "I'm not sure what I *do* think might happen."

"I believe there are a couple of concepts in the back of your mind about what might result from the stall, brought about by your idea of *too steep* a stall attitude."

Tom looked at the model again and tried to picture himself in the cockpit.

I went on. "I think that you see your plane either sliding downhill tail first, or flopping over on its back, when the stall breaks. Both concepts are enough to make the stoutest heart falter."

Tom's head started nodding with understanding and agreement.

"Of course, neither of those things will happen," I said.

"Why not?"

"Because your plane just isn't tilted up that high. Would you believe that your plane's nose is tilted only about 13 or 14 degrees when the stall breaks?"

"That's pretty hard to believe. It sure *seems* a heck of a lot steeper when I'm doing them."

"That's because of your myopic viewpoint when the plane stalls. Look at it this way—where are you looking when the stall breaks?"

"Right out over the nose."

"Exactly! And you see only . . . "

"Sky," he finished.

"And it *looks* like you're pointed straight up into outer space, because you don't see any points of reference. If you want to see your *true* stall attitude, look out at your wingtip. You'd see the slant of the wingtip relative to the horizon—way less than 20 degrees."

"Even power-on stalls?"

"Only about 3 degrees more in most planes," I said. "But in no way will you be steep enough to produce any imagined tail-slide or back-flop."

"What *does* the plane do when it stalls?"

"It settles about 20 feet in the stall attitude—trying to land—then the nose eases downward."

Tom looked at the model plane he had put on my desk. He pushed the tail down slightly with his finger, trying to mentally reduce his 50-degree pitch attitude concept to the 13 or 14 degrees I'd promised him.

"Tom," I said, "if you want to take an unhurried look at the stall attitude, just look at any parked tailwheel airplane," pointing to the little J-3 parked in the hangar. "We land 'em in a full stall, and *that's* the true stall attitude" (Fig. 1-20).

Fig. 1-20. Light-plane tailwheelers like this J-3 are designed to land full-stall in the three-point attitude. Feel free to take a long look at the true stall attitude. I sincerely hope that, sometime during your training, you find the chance to fly a tailwheeler. It's great fun. And, hey, if you *do* fly one, drop me a postcard to let me know what you thought of it.

"I hadn't thought of it that way."

"Well," I said, "when you realize how tame the stall attitude really is, you stop worrying about those things that won't happen."

"I think I'll take the Cherokee up for a few more minutes," he said as we walked toward the open hangar door into the bright mid-morning sunshine.

Tom walked across the ramp toward his plane, and I returned to my beat-up desk, sat down, looked at the fuel bills, glanced up at my rental rates posted on the wall, and tried once more to fit the square peg in the round hole.

■ **Flight at Minimum Controllable Airspeed**

Once you have a clear understanding of stall recovery techniques, the flight-test examiner will ask you to fly the airplane at *minimum controllable airspeed.* This is the ultimate maneuver with which to develop a keen sense of stall avoidance, because at *minimum controllable airspeed,* you must fly at an attitude and airspeed such that any reduction of power or airspeed, or any increase in angle of attack or angle of bank produces the indications of a stall.

When flying this maneuver, you are operating on a "razor's edge." And you should be able to demonstrate your ability to operate safely on this edge while performing the four basic maneuvers of (1) straight and level flight, (2) turns, (3) climbs, and (4) descents.

Straight and Level

When flying straight and level at minimum controllable airspeed, establish an airspeed close enough to stall speed to keep the stall-warning indicator honking or winking. You will find yourself flying well within the *region of reverse command* (Fig. 1-21), the point at which it takes

AIRSPEED FLOWN	POWER NEEDED TO MAINTAIN LEVEL FLIGHT
90 Knots	2400 RPM
85 Knots	2300 RPM
80 Knots	2200 RPM
75 Knots	2000 RPM
60 Knots	2400 RPM

Fig. 1-21. This is an example in which region of reverse command, the point at which it takes *more* power to fly *slower,* occurs at 60 knots. Near-cruise power is needed to maintain altitude close to the plane's minimum controllable airspeed.

more power to fly slower. Near-cruise power is required and the nose is pitched higher than the normal climb attitude. Expect a bit of slop in aileron control. The airflow over those control surfaces is diminished considerably, and its takes a lot of control deflection to make them do their job. Rudder control, of course, remains firm with the vertical stabilizer safely within the airflow produced by the propwash.

And it is well that rudder control remains firm. At minimum controllable airspeed, you are flying with airspeed well below cruise and with power at or above cruise. In this configuration, the left-turning tendency of "torque" is quite pronounced. You need firm right rudder to keep the plane flying straight. Let's take a moment to discuss torque.

Torque

Torque is the force that is always trying to turn the plane to the left anytime the plane's engine is running. This force is really made up of four *separate* forces, but since they all derive from the action of the propeller and all work to produce a left-turning tendency, they are lumped together simply as torque.

We will discuss two of the four forces—slipstream effect and propeller torque—now. The other two forces—P-factor and gyroscopic precession—will be discussed in chapter 3 concerning takeoffs.

SLIPSTREAM EFFECT. As the propeller whirls, it moves the air backward across the airplane. This blast of air is called the propwash or slipstream. The slipstream tries to turn the airplane left. The propeller, turning clockwise (from the pilot's viewpoint), corkscrews the air downward around the fuselage, then upward to strike the left side of the vertical stabilizer (Fig. 1-22).

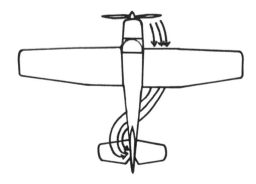

Fig. 1-22. Slipstream effect tries to veer the plane left.

PROPELLER TORQUE. In seventh grade science class you learned that for every action there is an opposite and equal reaction. And since the propeller is whirling clockwise (to the right), a recoil is generated, twisting the plane to the left. The pilot could hold right-rudder pressure throughout the flight to compensate for this left-turning tendency, but that would quickly produce a mighty tired right foot. With this in mind, the aircraft manufacturers have designed the planes to automatically compensate for torque.

COMPENSATING FOR TORQUE. Since the forces of torque stem from the propeller's movement, it stands to reason that the magnitude of torque varies with the speed of the propeller. Therefore, the manufacturer designs the plane to compensate for only *one* magnitude of torque, the one developed at cruise power and airspeed—the flight condition encountered the majority of time.

The manufacturer employs three torque compensators as the plane is built: canted engine installation; offset vertical stabilizer; and wingtip washout (Fig. 1-23). As you can readily see, each of these produces a counter, right-turning force. These are balanced to produce a straight-ahead flight at cruise power and airspeed.

This means the pilot must further compensate with rudder anytime other than when cruise power and airspeed are being used. In a high-

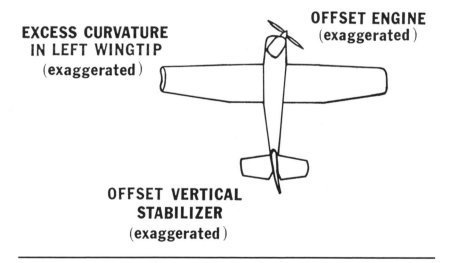

EXCESS CURVATURE IN LEFT WINGTIP (exaggerated)

OFFSET ENGINE (exaggerated)

OFFSET VERTICAL STABILIZER (exaggerated)

Fig. 1-23. Design features compensate for the left-turning forces of torque.

power climb or flight at minimum controllable airspeed, for example, the built-in compensations are not adequate to maintain straight-ahead flight—the plane tries to veer left. The pilot must temporarily apply additional right rudder. On the other hand, the build-in compensations are too great for a low-power glide. Thus the plane tries to veer to the right, and the pilot must apply *left* rudder to keep the plane traveling straight ahead.

TURNS. Turning at minimum speed is an increased challenge. The load factor increases as the wings are banked, which means simply that stall speed increases in the turn. Since our minimum controllable airspeed is keyed to stall speed, it means that minimum speed also increases in the turn. And the greater the angle of bank, the greater this minimum controllable speed becomes. (In a typical light trainer, this increase is about 4 or 5 knots in a 30 degree bank.)

It becomes clear, then, that you need to increase your flying speed in a turn. You can lower the nose to pick up speed, but this will cost altitude, which you want to maintain. The most practical way to gain the extra speed is with extra thrust. Advance the throttle about 200 rpm as you enter the turn, and withdraw those extra rpms as you roll out of the turn. Your altitude should stay "right on the money," and you will have learned an important lesson: Keep your right hand on the throttle anytime nimble action may be needed in the cockpit.

Coordination between aileron and rudder control is complicated in turns at minimum controllable airspeed. With torque acting like left rudder, executing turns to the right and to the left is dissimilar. In entering the left turn, for example, very little left rudder is needed to initiate the turn—torque is doing most of that job for you. During the left turn, the "ball" (Fig. 1-24) will tell you to keep some pressure on right rudder, for torque is still in play, just as it was in straight and level minimum-

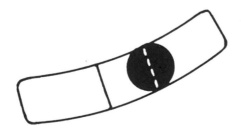

Fig. 1-24. The ball, deflected to the right, calls for right-rudder pressure, or less left aileron.

speed flight. When rolling out from a left turn, expect to need considerable right rudder to initiate the recovery. Here you need enough rudder to both eliminate the bank *and* control torque.

As you suspect, a right turn at minimum speed calls for different rudder action during roll-in and roll-out. It takes firm right rudder to slant those wings into a bank away from the force of torque. On roll-out, however, little left rudder is called for because torque is helping out. *During* the turns, however, rudder action is similar in both directions; hold just enough right rudder to counter torque and just enough to keep the ball caged.

CLIMBS. When climbing at minimum controllable airspeed, expect to hold *considerable* right rudder, to keep the plane flying straight ahead. You will be flying under full climb power with an airspeed below either best-rate or best-angle-of-climb speed. Strong torque will be in effect.

Since you are climbing at an airspeed well within the region of reverse command, don't expect much performance from the plane; it will climb with all the robustness of a rusty jackscrew. For this reason, confine these climbs to 100 or 200 feet—the engine may start to overheat by that time anyway.

DESCENT. To descend at minimum controllable airspeed, just use a slightly lower power setting than was required to maintain straight and level minimum-speed flight. Since your flight path is now downward, you also need to relax a bit of back pressure to maintain the same angle of attack.

Flaps

Once you have mastered the four fundamentals at minimum controllable airspeed, flying the maneuver with flaps should pose no problem. In doing so, you will learn two facts about flaps that are important for your up-coming landing demonstrations. First, you will find that partial flaps lower the stall speed (and minimum controllable airspeed) about 3 to 5 knots in the average trainer. You will also find that increasing the flap extension to "full" gives little further reduction in stall speed, because partial flaps provide additional lift, while full flaps provide mainly drag.

Second, you will discover that flaps increase your forward visibility. They allow you to fly slowly at a much lower pitch attitude—the cowl no longer covers the horizon. This increase in visibility is most welcome during those last few moments on the final approach to a landing.

Flight-Test Objectives

Flight at minimum controllable airspeed is an exacting maneuver that hones a pilot's awareness of aircraft performance and which leaves few reserves.

You have a two-fold objective in mastering this maneuver. In addition to precision flying, you want to prevent a stall. Of course, you are flying deep within the stall indicator's warning. The first indication of an impending stall will probably be either the shuddering "buffet" or a rapid further loss of aileron.

If either warning appears, take quick action. Put your stall recovery into play by reducing your angle of attack, adding power, and leveling the wings. When you reduce the angle of attack, do so by lowering the nose only slightly. Remember that you are not trying to regain a civilized flying speed, you just want to keep flying on the edge. And when adding power, take care not to let the nose ride up—you just don't have any angle of attack in reserve.

As your flight test draws near and your skill is at its peak, go ahead and play with the maneuver. Intentionally induce a stall. Back the throttle down 200 rpm to lose the needed thrust, and watch the stall happen. Or bank the wings to pick up aerodynamic loading, and watch the stall happen. Or nudge the nose up a degree or two toward the critical angle, and again watch the stall happen.

Once you master this maneuver you are entitled to a slap on the back, and you can start thinking of yourself as a pilot. You are also ready to begin the next phase of the flight test—performing the ground reference maneuvers.

IN REVIEW

■ A stall occurs when there is an insufficient flow of air passing over an adequate area of wing surface.

■ An insufficient airspeed cannot provide an adequate flow of air.

■ An excessive angle of attack diminishes the available lifting area of the wing.

■ Three factors control the airspeed at which a wing stalls:

1. Density of the air
2. Aircraft weight
3. Engine power

The air density is controlled by three factors:

1. Altitude
2. Temperature
3. Moisture content

An increase in aircraft weight calls for a greater airflow with which to provide the extra lift required.
Increase the plane's weight (and stall speed) through two actions:

1. Increase the payload
2. Impose aerodynamic loading

When under a lot of power, the propeller acts like a fan blowing a mass of fast-moving air across the wing.
Increasing the wing's angle of attack reduces the lifting surface of the wing by allowing the separation point to move forward.
Each wing design has a critical angle of attack beyond which lift erodes.
A critical angle of attack remains virtually the same at any airspeed, aircraft attitude, or loading.
The critical angle of attack in most light planes occurs between 12 and 17 degrees.
Extended flaps allow you to enjoy an advantageous angle of attack, while not allowing the separation point to creep forward.
Control coordination plays a significant part in a wing's stall characteristics.
Design features in light-plane wings help produce a "mellow" stall:

1. Wingtip washout
2. Wing shape
3. Wing slots
4. Stall strips

Pilot distraction is the major source of accidental stalls.
Recover from a stall by taking four actions:

1. Reduce the angle of attack
2. Add power
3. Level the wings
4. Regain any altitude lost

Many pilots who maintain anxiety toward stalls do so because of an exaggerated concept of the stall attitude.

The principal objective of flight at minimum controllable airspeed is to teach stall recognition and recovery.

Aileron control diminishes at minimum speed — rudder control remains responsive.

The left-turning tendency of torque is present throughout flight at minimum controllable airspeed.

The minimum controllable airspeed increases in a turn due to imposed load factors.

■ Additional power is required in a turn to maintain minimum controllable airspeed.

Control coordination during turns at minimum controllable airspeed is complicated by torque.

Restrict minimum controllable airspeed climbs to 200 feet to reduce engine overheating.

FLIGHT-TEST GUIDELINES

Maneuvering at Critically Slow Airspeed

Imminent Stalls

An imminent stall means that if you do not take immediate action, a full stall will occur. In asking you to demonstrate a recovery from an imminent stall, the flight examiner wants to determine if you can promptly recognize an oncoming stall and if you can effect a recovery to normal flight before the full stall develops. To make this determination, the examiner will ask for a demonstration both with and without the engine developing power.

POWER ON. This maneuver simulates a stall approaching after takeoff. When demonstrating recognition and recovery from power-on imminent stalls, perform the maneuver in six steps:

1. Climb to a safe demonstration altitude — a height of 2500 feet above ground level is a reasonable minimum safe altitude for most light training aircraft. Add another 1000 feet if you are flying a sleeker model. Remember, a spin can result from a stall, and these may take a *lot* of altitude for recovery. The *last* thing you want on board during a flight test is a nervous examiner.

2. Execute a 90-degree clearing turn. During the demonstration, the high-riding cowl masks a lot of sky and there could be conflicting

traffic. So, take a hard look around, then turn the plane 90 degrees for another search of the sky. Again, this action keeps that examiner relaxed. Every examiner that I know makes a solemn oath to their kids when they leave for work each morning; they promise not to bump into another plane. (Remember to also look for planes that might pass *beneath* you.)

3. Point the plane along a definite entry heading and make a mental note of the heading and the entry altitude.

4. Pull the plane into a best-angle-of-climb speed, with climb power. Keep the ball centered using rudder against torque, as in any climb. This control coordination prevents a "skid," which would create an uncertain flight path and stall characteristics.

5. As the airspeed needle reaches best-angle-of-climb speed, continue increasing the angle of attack until you notice one of the warnings that a stall is imminent:

   a. Aircraft buffet.
   b. Sudden decay in aileron control.

   These indications will normally occur *after* the plane's stall warning has activated.

6. Effect an immediate five-step recovery at the first indication of an imminent stall:

   a. Reduce the angle of attack to get the wing safely flying again.
   b. Add any power still available to increase the windflow over the wing.
   c. Stop any turn using rudder and ailerons to prevent any extra aerodynamic loading.
   d. Return the plane to its entry heading.
   e. If the recovery has the plane below the entry altitude, climb back to that height.

POWER OFF. Recovery from an imminent stall at idle power simulates the action to take when you find yourself flying dangerously slow during the approach to landing. Demonstrate your ability to recover in seven steps:

1. Climb to the same safe entry altitude used for power-on stalls.
2. Execute a 90-degree turn to clear for conflicting traffic.
3. Establish a definite entry heading and altitude. Make a mental note of them.

4. Apply carburetor heat, throttle down to idle power, and try (unsuccessfully, I might add) to maintain your existing altitude by steadily increasing the angle of attack.
5. When the airspeed indicator slows toward normal approach speed, extend the flaps to simulate the landing configuration.
6. Keep increasing the angle of attack until you recognize either indication of an imminent stall:

   a. Aircraft buffet.
   b. Sudden decay in aileron control.

7. Demonstrate an immediate recovery:

   a. Get the wings safely flying with a reduced angle of attack.
   b. Increase the windflow across the wing by applying full throttle.
   c. Prevent aerodynamic loading by leveling the wings.
   d. Return the plane to the maneuver's entry heading and climb at best-rate airspeed.
   e. Once climbing at the proper airspeed, retract the flaps one notch at a time.
   f. Level off on the entry altitude and re-establish cruise flight.

FLIGHT-TEST TOLERANCES
1. Proper control coordination throughout the demonstration.
2. Recognize first indication of imminent stall and effect an immediate recovery.
3. Prevent a full stall from developing.
4. Avoid an excessive loss of altitude.
5. Be able to explain the aerodynamic elements of an imminent stall.

(Note: The examiner may ask you to demonstrate stall recognition and recovery while flying through a shallow bank. This action simulates an inadvertent, dangerously low airspeed while departing the pattern or turning from the base leg to final approach. No problem, as long as coordinated flight is maintained—stall recognition, characteristics, and recovery remain unchanged.)

Full Stall—Power On

Forcing a full stall means that you have deliberately flown your plane past the point of "imminent stall." By the time you reach the full-stall condition, all the plane's warnings have been delivered: the stall warning device has activated, you have felt the aileron decay, and you have felt the buffet. The indication of a *full* stall is singular in nature — the nose pitches downward from the flight path. To many, this pitching *seems* excessive, but it is not. The nose pitches only a few degrees below the level-flight attitude, trying its best to make the wings fly again with a reduced angle of attack. (If the pilot, however, *insists* on maintaining back pressure on the stick, the wings don't get a chance to regain their flying status. Then the nose *will* steepen its descent as it naturally follows the plane's center of gravity, which lies ahead of the wing's center of lift. At the point of full stall, of course, the pilot must *help* the plane seek a satisfactory angle of attack by eliminating the back pressure on the stick.)

In asking you to demonstrate stall recognition and recovery from power-on full stalls, the examiner wants to see how you would handle an inadvertent full stall shortly after takeoff. The examiner wants to see a prompt and correct technique that results in a minimum loss of altitude.

You may be asked to demonstrate a power-on recovery while turning in a shallow bank that simulates the first climbing turn away from the runway. As long as you make a coordinated turn, the stall character and recovery remains unchanged. If, however, you are skidding when the stall occurs, recovery is complicated. A sharper stall may occur, with a heightened potential to develop into a spin. A delay in recovery time is created, for the pilot must take a moment to get the controls coordinated once more. An excessive loss of altitude is a certainty — and a large altitude loss is something the examiner does not want to see. This extra loss of altitude doesn't amount to a "hill of beans" when it occurs at a safe demonstration height, but at pattern altitude with marginal safety below, it is another matter. (Note: Power-on stalls will normally break a little sharper than those entered with only idle power. This is natural. Full power develops a propwash that keeps the wing roots flying nearly as long as the wingtips. When the stall occurs, almost the entire wing stalls at once. Do not hesitate to convey your knowledge of this factor to the examiner.)

When asked to demonstrate recognition and recovery for power-on full stalls on the flight test, perform your maneuver as a series of seven steps:

1. Climb to a safe demonstration altitude.
2. Clear for traffic with a 90-degree turn.
3. Establish a definite entry heading and altitude, and make a mental note of them.
4. Establish a best-angle climb with appropriate climb power.

5. Continue to increase the angle of attack until the nose dips, signifying a full stall.
6. Initiate a recovery before the nose descends through the horizon:

   a. Decrease angle of attack.
   b. Add all available power.
   c. Level the wings.

7. Re-establish the entry heading and climb to replace any altitude lost.

FLIGHT-TEST TOLERANCES

1. Proper control coordination throughout demonstration.
2. Recognize full-stall indication.
3. Effect prompt recovery that allows minimum loss of altitude.
4. Avoid a secondary stall.
5. Be able to explain the aerodynamic factors associated with a power-on stall. Display an awareness of flight situations that produce stalls.

Full Stall—Power Off

Your demonstration of a recovery from a full stall with power at idle shows your ability to safely handle an inadvertent stall during the landing approach. Therefore, enter the stall with the plane in its landing configuration—flaps and gear extended. Remember, when you add power during the recovery, the nose will try to pitch upward to further *increase* the angle of attack. Many applicants overlook this, and thereby create a secondary stall.

The examiner may ask you to stall in a medium-banked turn that simulates an approaching pilot turning from base to final. Again, a turn adds no complexity to the recovery as long as the controls remain coordinated. Uncoordinated controls, however, seriously affect your satisfactory recovery—they will delay recovery and cost an excess loss of altitude. And remember that this recovery demonstration simulates action *close* to the ground.

When asked to perform this maneuver, plan a demonstration that follows nine steps:

1. Climb to a safe demonstration altitude.
2. Look for conflicting traffic with a 90-degree turn.
3. Open the carburetor heat.
4. Establish a definite entry heading and altitude. Make a mental note of these values.

5. Idle the engine and decelerate to approach speed.
6. Keep increasing the angle of attack while you maintain any turn assigned.
7. Increase the angle of attack until the nose pitches forward, indicating a full stall.
8. Take immediate action to recover from the stall:

   a. Decrease the angle of attack.
   b. Add all available power.
   c. Stop any turn with coordinated ailerons and rudder.
   d. Establish a best-rate climb while returning to the entry heading and altitude. Retract the gear and flaps (one increment at a time) while you do so.

9. Re-establish normal cruise speed at entry altitude and heading.

FLIGHT-TEST TOLERANCES

1. Proper control coordination throughout demonstration.
2. Maintain a 30-degree banked turn (+/− 10 degrees) if assigned.
3. Immediately recognize a full stall and initiate a prompt recovery with minimum loss of altitude.
4. Effect a smooth flap retraction.
5. Be able to explain the aerodynamic factors associated with the power-off stall.
6. Display an awareness of any flight situation that may lead to an inadvertent stall.

■ Maneuvering at Minimum Controllable Airspeed

Minimum airspeed, here, means that any increase in the angle of attack or bank or any reduction in power or speed will result in an imminent stall:

Increased Angle of Attack

You are on the razor's edge of flying. Any increase in the angle of attack penetrates the wing's critical angle, unless you take compensating action.

Increased Angle of Bank

Even the slightest increase in the G load resulting from an increased angle of bank pushes you "over the edge" unless you take compensating measures.

Decreased Airspeed

At minimum controllable airspeed, the wings need every knot of windflow that they possess just to stay aloft. You must employ compensating measures anytime you maneuver away from straight and level.

As you perform the requested minimum airspeed maneuvers, you have a threefold job at the controls:

1. Operate the plane safely.
2. Prevent the occurrence of an imminent stall as you fly the maneuver.
3. Perform the maneuver within prescribed tolerances of accuracy.

Straight and Level Flight at Minimum Controllable Airspeed

Fly the maneuver in a series of steps:

1. Climb to a safe demonstration altitude—a height of 2500 above ground level (AGL) is a reasonable, safe altitude for most light training aircraft when performing the minimum-airspeed maneuvers. Add another 1000 feet if you are flying a higher-performance airplane. Remember, an error on your part could result in a stall, which may turn into a spin. Provide ample altitude to assure a safe recovery.
2. Execute a 90-degree clearing turn. During these minimum-airspeed maneuvers, much of your forward visibility is obstructed by the high-pointing cowl. Before entry, you must make certain that your section of the sky is safe from traffic. (And *during* the maneuver, remember to look out each side window to make sure the sky you *can* see remains clear of traffic.)
3. Point the plane along a definite entry heading, by using either the heading indicator or a reference line on the ground. Make a mental note of the heading and entry altitude.
4. Establish minimum controllable airspeed. First, throttle down to 2000 rpm. As the speed bleeds away, start lifting the nose slowly to maintain altitude. Keep increasing pitch until you feel the first indication of an imminent stall at your reduced power setting—quickly note the airspeed. The airspeed that produces an imminent stall at 2000 rpm will be your minimum controllable airspeed at advanced power. Quickly advance power to hold altitude at the slow airspeed. Don't be surprised if it takes near-cruise power to do so; you are flying deep into the *region of reverse command,* where it takes *more* power to fly *slower.*

   Once your power is adjusted, trim the plane to the attitude that maintains minimum controllable airspeed. The left-turning tendency of torque is pronounced in this configuration of pitch, airspeed, and power. Firm right rudder is needed to hold the entry heading. (If your plane has rudder trim available, use it.) Your task now, is simply to maintain minimum airspeed, altitude, and heading, while keeping the controls coordinated throughout the maneuver.

5. The examiner may ask you to extend the flaps while maintaining minimum airspeed. To prevent large pitching moments, extend the flaps one increment at a time—each increment of extension will require a slight pitch reduction to keep the airspeed captured. Retrim with each flap extension for positive airspeed control. You will find that the final flap setting produces far more drag than additional lift. Add power to counter this extra drag to maintain airspeed. You may find that near-climb power is required to do the job. A quick reference to the vertical airspeed indicator (VSI) tells the story. If the VSI indicates a downward trend, you are sinking; add power. If the VSI shows a climb, reduce throttle to maintain the entry airspeed while holding altitude.

6. To recover from the maneuver and reverse the process:

   a. Retract the flaps by increment, letting the airspeed increase.
   b. As airspeed increases, slowly return pitch and trim to its cruise attitude.
   c. Once cruise speed is attained, readjust the throttle to cruise power, and retrim.

FLIGHT-TEST TOLERANCES

1. Proper control coordination throughout demonstration.
2. Prevent the occurrence of an imminent stall throughout demonstration.
3. Maintain the entry altitude +/− 100 feet.
4. Maintain the entry heading +/− 10 degrees.
5. Maintain minimum controllable airspeed within 5 knots.

Turns at Minimum Controllable Airspeed

Again fly the maneuver as a series of steps:

1. Establish a safe demonstration altitude.
2. Look for conflicting traffic with a 90-degree clearing turn.
3. Establish a definite entry heading and altitude. Make a mental note of these values.
4. Establish minimum controllable airspeed while maintaining heading and altitude.

5. Initiate a 360-degree turn in the requested direction, at the bank specified (15 to 30 degrees).
6. Maintain altitude throughout the turn. After establishing the desired bank, add slight power to compensate for the additional load factor.
7. Maintain coordinated controls throughout the turn. At the low airspeed and high angle of pitch, torque will give the effect of applied left rudder. This means that when turning to the right, considerable right rudder is needed to establish a coordinated turn. When turning left, on the other hand, very little left rudder is required.
8. The examiner may call for extended flaps prior to, or during, the turn. Remember to extend them by increments, adjusting trim and pitch with each extension. The extension of *full* flaps may require near-climb power to maintain airspeed against the additional drag produced by the flaps. This power setting further intensifies the left-turning tendency of torque, calling for additional rudder action on your part.
9. Recover from the turn on the entry heading. If your direction of turn was to the left (with the torque behind it), brisk right rudder is needed on the roll-out if the controls are to remain coordinated. A roll-out from a right turn, on the other hand, requires a little left rudder. Maintain your altitude during the roll-out. Reduce power to the straight and level minimum speed as you return the wings to level.
10. If you fly the turn with extended flaps, retract them after recovering to straight and level minimum airspeed flight. To maintain altitude while doing so, retract them one increment at a time, adjusting trim and pitch with each increment of retraction.
11. Return the plane to normal straight and level cruise flight to complete the maneuver. Keep the controls coordinated and maintain entry heading and altitude while you do so. To keep the controls coordinated while maintaining your heading, you must compensate for diminishing torque during the recovery. Begin the recovery with firm right rudder held against significant torque—as your plane accelerates toward cruise speed, this torque diminishes, so must your right-rudder pressure.

   Your best chance at maintaining altitude lies within your smoothness with throttle and stick. Smoothly restore cruise power, then slowly reduce pitch as the airspeed climbs to cruise. A smooth touch with controls should have you hitting cruise speed just as your nose levels.

FLIGHT-TEST TOLERANCES

1. Proper control coordination throughout the maneuver.
2. Prevent the occurrence of an imminent stall.
3. Maintain the specified bank angle within 10 degrees.
4. Roll out within 10 degrees of the specified recovery heading.
5. Maintain airspeed within 5 knots, altitude within 100 feet.

Climbs at Minimum Controllable Airspeed

Fly your demonstration climb as a series of steps:

1. Establish a safe demonstration altitude.
2. Look for conflicting traffic with a 90-degree clearing turn.
3. Establish an entry heading and determine your desired recovery altitude.
4. Establish minimum controllable airspeed in straight and level flight.
5. Initiate a climb toward your desired recovery altitude. Advance the throttle to climb power and slowly increase the pitch necessary to maintain your established minimum controllable airspeed. Don't expect much climb performance—you are deep into the region of reverse command. The plane will move upward with all the pep of a rain-soaked blimp.
6. Maintain your entry heading throughout the climb. Torque produces a pronounced left-turning tendency, and firm right rudder is needed to keep your climb on track.
7. Begin your level-off procedure slightly below your desired recovery altitude; you will gain extra yards of height as you level out. As a rule of thumb, begin to slowly lower your nose about 10 feet below your desired altitude for each 100 feet per minute registered on the VSI during the climb.
8. Level out at cruise speed to complete the maneuver. Time your reduction in pitch so that the nose levels just as you reach the desired altitude. Leave the throttle at climb power until the plane reaches cruise speed. Then, promptly reset the throttle to cruise power. If you delay power reduction, you will climb several feet through your desired altitude. If you reduce power *before* you reach cruise speed, you will settle below your planned recovery altitude.

Note: Your examiner may call for this demonstration with flaps extended. If so:

   a. Extend flaps one increment at a time while maintaining straight and level flight at minimum controllable airspeed, before you commence the climb.

    b. Expect only the slightest rate of climb during the ascent.

    c. Retract the flaps by increments, once leveled at minimum controllable airspeed.

FLIGHT-TEST TOLERANCES

1. Proper control coordination throughout the maneuver.
2. Prevent the occurrence of an imminent stall.
3. Maintain the entry heading within 10 degrees throughout the maneuver.
4. Maintain minimum controllable airspeed within 5 knots.
5. Level off within 100 feet of the specified recovery altitude.

Descents at Minimum Controllable Airspeed

Perform your demonstration of a descent as a series of steps:

1. Establish a safe demonstration altitude. An entry altitude 2500 feet above the specified recovery altitude is reasonable for most light training aircraft.
2. Look for conflicting traffic with a 90-degree clearing turn.
3. Open the carburetor heat. The descent is made under reduced power and carburetor ice is possible.
4. Establish straight and level flight at minimum controllable airspeed on an entry heading and altitude. Make a mental note of these values.
5. Initiate the descent. Begin by reducing power. If flying a fixed-prop plane, reduce power by 100 rpm for each 200 feet per minute (fpm) you wish to achieve on the VSI. (If flying a controllable prop, change this power value to 1″ Hg per 200 fpm.)

    As the descent begins, do one other thing in order to prevent accelerating through minimum controllable airspeed. As you start going downhill, lift the plane's nose a "tad"—about a half-dot on the attitude indicator for each 200 fpm of descent. If you fail to do this, the plane will pick up speed. There is an interesting relationship between pitch, power, and airspeed that exists in most light airplanes.

    a. One-half dot on the attitude indicator = 200 fpm on the VSI = 5 knots on the airspeed indicator.

    b. 100 rpm (or 1″ Hg) = 5 knots = one-half dot on the attitude indicator.

    A few examples of this relationship at play demonstrate the concept:

      ■ If you reduce power by 100 rpm, you must adjust pitch by a one-half dot deflection to maintain altitude and you will slow by about 5 knots.

■ If you lower the nose by one-half dot and adjust the throttle to maintain rpms, you will descend about 200 rpm and pick up 5 knots of airspeed.

■ If you lower the nose by one-half dot and leave the throttle unchanged, the rpms will increase by 100 as you descend about 200 fpm. Your airspeed in this instance will increase 10 knots— 5 knots for the 200 fpm and 5 knots for the extra 100 rpm.

6. Recover from the descent. Lead this recovery by 20 feet above the desired recovery altitude for each 100 fpm rate of descent. With this lead, begin returning to level minimum controllable airspeed power and start repitching the nose to its level minimum attitude. The plane will continue to descend during these moments of lead and you should hit your recovery altitude right on the mark.

7. Recover to straight and level at normal cruise. The examiner may have you fly the descent with extended flaps. If so, retract them one increment at a time while in straight and level minimum airspeed flight before you restore normal cruise power.

FLIGHT-TEST TOLERANCES

1. Proper control coordination throughout the maneuver.
2. Prevent the occurrence of an imminent stall.
3. Maintain entry heading within 10 degrees; maintain minimum controllable airspeed within 5 knots; level off within 100 feet of the specified recovery altitude.

Combined Minimum Controllable Airspeed Maneuvers (Turning Climbs and Descents)

The examiner may request that you demonstrate a climbing or descending turn to a specified recovery altitude and heading.

1. Simply combine the techniques of the basic minimum controllable airspeed maneuvers.
2. Apply compensating rudder action to keep the controls coordinated during changes in pitch, power, and bank.
3. You may not be able to predict which recovery goal you will achieve first—altitude or heading.
4. If you gain the specified altitude before you reach the recovery heading, level at minimum controllable airspeed while you continue the turn.

5. If you reach the specified heading first, roll out of the turn while you continue the climb or descent to the recovery heading.
6. If flaps are to be employed, extend them by increments while maintaining straight and level minimum airspeed flight before you initiate the maneuver. Retract the flaps by increment after you have re-established straight and level flight at minimum controllable airspeed.

Flight at minimum controllable airspeed exercises nearly every function of basic airmanship, from precise airspeed control to control coordination to power management. Your satisfactory demonstration of the maneuvers displays your mastery of the aircraft. You, in fact, demonstrate your total awareness of self, machine, and environment as you perform with precision.

2.

Ground Reference Maneuvers

Pilots are true riders of the wind, who spend their lives aloft with the capricious wind—sometimes with its favor, often with its malice. They share this force of nature with all things that fly—the bird carrying a promise of song, the missile with its shriek of a thousand individual horrors, the ball with the hope of a first down. The wind exerts significant influences on your plane and on your success with the flight test. These influences are most visible in matters of *groundspeed* and *ground track*.

■ Groundspeed

Wind is, of course, just air in motion. You have to deal with air in *vertical* motion when you hold the wingtips steady in the rippling air (turbulence) of straight and level flight. Air in *horizontal* motion really does little to influence the aerodynamic flight of the plane, but it does influence the speed at which the plane covers the ground.

Headwinds and Tailwinds

If your plane is flying at 100 knots (airspeed) into a direct headwind of 10 knots, it covers the terrain at 90 knots (groundspeed) (Fig. 2-1). If

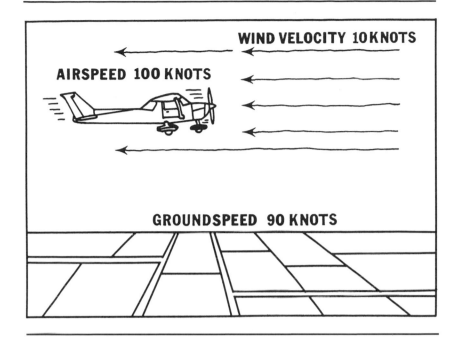

Fig. 2-1. Wind influences groundspeed.

the same plane was experiencing the same windspeed from the tail, groundspeed would be 110 knots.

We speak of headwinds or tailwinds in terms of *wind components;* by this we mean the direct wind equivalent of an angling wind. For example, a 10-knot wind coming 40 degrees off the nose produces a *headwind component* of 8 knots. The same wind from 60 degrees produces a component of only 5 knots, and the same wind from 90 degrees produces no headwind component at all. Wind components are useful when figuring takeoff and landing distances, as well as when planning the time enroute required for a cross-country flight.

Crosswinds

Just as we speak of headwind and tailwind components, we also speak of *crosswind components.* The crosswind component is the force that an angling wind delivers, expressed as a direct (90-degree) crosswind value. A 10-knot wind that is 40 degrees off the nose, for instance, causes the plane to drift with a 6-knot crosswind component, or with a force equal to a 6-knot direct crosswind.

Interestingly, the slower the plane flies, the more the wind affects ground track. If two planes, for example, fly through the same crosswind, but at different flying speeds—the first at 60 knots, the second at 100 knots—for any given distance flown, the wind will try to drift the slower plane much farther off the desired ground track than it will the faster plane. This is simply because for any given distance, the wind has more time to work against the slower plane. Therefore, the slower the groundspeed, the greater the crosswind correction must be, if you are to maintain a desired ground track. Understanding this simple relationship between groundspeed and wind drift is critical to correctly performing ground reference maneuvers in the flight test.

■ Ground Track

If an airplane flying on a windless day crosses the terrain in the same direction that the plane's nose is pointed, *ground track* and *aircraft heading* are equal. The same happens when the plane flies against a direct headwind or with a direct tailwind—but a totally windless day is rare, as are direct headwinds and tailwinds. The wind usually crosses your path at an angle. Then, unless you take positive action, the wind drifts the plane from the desired ground track (Fig. 2-2).

This wind drift can be significant. Consider flying 250 miles from Miami, northward, to visit Aunt Minnie in Jacksonville, through an easterly wind aloft of only 20 knots. If you fail to correct for wind drift,

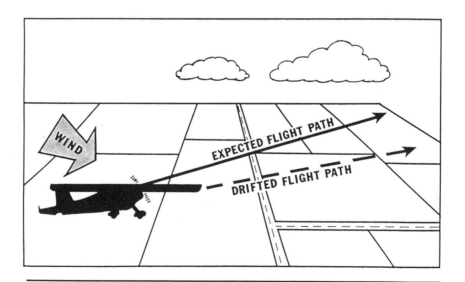

Fig. 2-2. Unless corrected for, wind influences ground track.

the landing might hold a surprise. When it is time to land, you might arrive 50 miles west, at the lovely rural airport of Lake City, with Aunt Minnie waiting in Jacksonville. Even immediate navigation becomes a problem if wind drift goes uncorrected. On your final approach, for example, if you do not correct for a moderate crosswind, you could miss the runway entirely and land in the grass. The flight-test ground reference maneuvers are designed to help the examiner evaluate your ability to cope with drift.

■ Evaluating Wind

When dealing with ground reference maneuvers, either in practice or during the flight test, you are concerned with the winds of the next 100 yards. Here, the wind's signposts on the ground below are the best indicators of wind direction and velocity. Inland waters, for example, give a reliable indication of the direction and velocity of surface winds (Fig. 2-3). If the wind is 5 knots or so, you see a glaze on the upwind shore; 10 knots add wind streaks to the water; and 15 knots produce whitecaps.

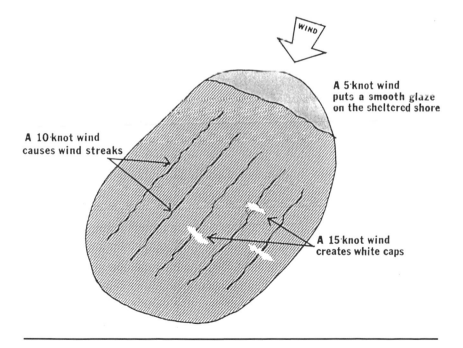

Fig. 2-3. Inland waters provide a good indication of the wind's direction and velocity.

Another indicator is smokestacks (Fig. 2-4)—5 knots bend the column of smoke slightly; 10 knots bend the stream of smoke at a 45-degree angle; and 15 knots bend the smoke almost parallel to the ground as it comes out of the stack. (My rural students tell me to look at the direction the cows are facing, but I can never remember whether they stick their noses or their tails into the wind. Besides, I've never seen all the cows facing the same way. I can only surmise that cows are notorious liars, and I disregard them.)

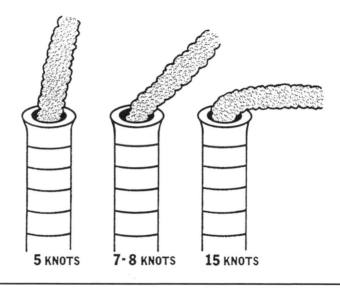

5 KNOTS   7·8 KNOTS   15 KNOTS

Fig. 2-4. Smoke from stacks has served as wind direction and velocity indicators for generations of pilots.

As you evaluate the surface wind indicators while performing ground reference maneuvers, keep two generalizations in mind. First, wind velocity usually increases with altitude. At the lower training altitudes, the wind's surface speed normally doubles at 2000 feet. Second, wind commonly shifts in a clockwise direction with an increase in altitude; you can normally expect a shift of 45 degrees to the right at 2000 feet.

So, with the basic facts of wind drift correction in hand, let's move on to ground reference maneuvers.

■ Correcting for Wind Drift

A pilot corrects for wind drift in one of two ways. One method is called *crabbing,* which is most often used when correcting for the wind over a long duration, such as during a cross-country flight. The other method is called *slipping.* This method is most often used when preventing drift over a shorter duration, such as those final moments of a landing approach. The flight test evaluates your skill with each method.

Crabbing

When you prevent drift by crabbing, you merely deflect the plane's nose a few degrees *into* the wind, as when a boat deflects its bow into the current when crossing a river.

When establishing the crab angle into the wind, simply make a coordinated turn of a few degrees. Once the wind correction angle is established, there is no need to hold rudder into the wind—neutralize the controls for normal straight and level flight. The nose is displaced only a few degrees from the desired ground track.

Slipping

When correcting for drift during those moments on final approach just before landing, use a slip. In slipping, use the ailerons to bank the wings toward the wind, while holding *opposite* rudder pressure. This will achieve two important goals during a landing. First, the banked wings divert the "tugging power" of lift into the wind, thus eliminating drift (Fig. 2-5). Obviously, you do not want to touch down on the runway

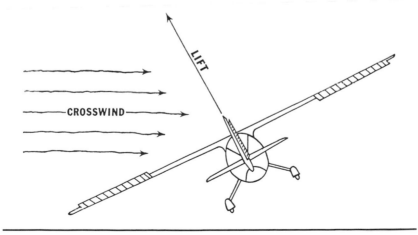

Fig. 2-5. The deflected lift of banked wings eliminates wind drift.

with the plane drifting sideways. Not only would this produce hard-to-manage directional control once on the ground, but the tires would screech with a howl that passengers would talk about for weeks.

Second, the slip holds just enough opposite rudder to keep the nose pointed straight ahead. This, too, promotes directional control at touchdown. Without a doubt, a plane that touches down at a slant to the runway is going to swerve—possibly hard enough to get a wingtip into the runway. Now, it is true that not many people are *hurt* by a tattered wingtip, but the cost of repairs can easily call for a second mortgage on the plane.

As you can imagine, slipping takes a bit of dexterity. To achieve this maneuver, you will have to utilize all the control coordination skills you have learned. You actually cross-control. But you must learn to slip if your flight-test landings are to be uneventful. And learning to slip is an important piloting skill that you develop during the ground reference maneuvers, where you learn to work with the wind.

AILERONS IN THE SLIP. Ailerons in the slip prevent wind drift, nothing more. By banking the wings into the wind, you divert lift toward the crosswind. This accomplishes the same effect as pointing a boat's bow into the river current to prevent drift. The tugging power of lift deflected windward holds the plane on course.

Just *how much* to bank the wings, of course, depends on the strength of the crosswind: a little wind, a little bank; a lot of wind, a lot of bank. A long, straight road below gives you the cue. If the slip you establish first still lets the plane drift, then you need to increase the bank; if the plane actually slides *toward* the wind, you obviously have to reduce the bank. Use the exact degree of bank that stops the plane's lateral movement.

Any banking you do with the ailerons must coordinate with rudder pressure.

RUDDER IN THE SLIP. In the slip the purpose of the rudder is simply to steer the nose, to keep the plane's fuselage straight in the flight path. If you banked the wings and did nothing with the rudder, the nose would quickly turn toward the bank. Therefore, apply *opposite* rudder immediately after banking the wings, to keep the nose pointed straight ahead.

How much opposite rudder do you use? Just as much as it takes to hold the nose straight. If you use a lot of aileron to correct for drift in a stiff breeze, then use a lot of opposing rudder; use a little aileron, then use a little rudder. Most pilots flying a slip tend to use insufficient rudder, because they are used to steering a car with their hands, not their feet.

AIRSPEED DURING THE SLIP.   One final concern during the slip. The plane wants to slow down because deflected rudder and ailerons produce excess drag. Lower the nose a bit to keep your approach speed right on target. (Airspeed indications are often inaccurate during a slip, due to the position of the static port. Learn to gauge approach speed by sound and feel.)

TRACKING ALONG A ROAD

I envy the butterfly. It spends hours aloft just fluttering with the wind, slurping some nectar here, enjoying a petunia there, passing pollen around wherever it lands. But pilots can't just float around with the breeze. To travel straight and true from point A to point B, they nearly always need to reckon with a wind.

In most of your travels aloft, you correct wind drift by crabbing. Learning to crab and performing it on the flight test is simplicity itself. First, just find yourself a long stretch of straight road, as nearly square with the wind as you can find. Next, descend to an altitude that makes wind drift apparent—about 1000 feet above ground level (AGL) works fine. Then fly along the right side of that road (to keep it in sight) at cruise speed.

Once alongside the road, establish a trial wind correction or crab angle. To decide how much crab is needed, you must first estimate the velocity and direction of the surface wind. (Remember, at 1000 feet you can expect a 50% increase in velocity and a shift in direction of about 20 degrees to the right.) Then, figure the crosswind component with a rule of thumb:

Angle of Wind to Nose or Tail	Crosswind Component
30 degrees	½ wind velocity
45 degrees	⅔ wind velocity
60 degrees	¾ wind velocity
90 degrees	Full wind velocity

At cruise speed in a typical trainer, plan a crab angle of about 1 degree for each 2 knots of crosswind component. For example, a plane cruising at about 90–100 knots would need a crab angle of approximately 10 degrees with a crosswind component of 20 knots. After you establish a trial crab angle, look for any remaining drift, then subtract or add 2 or 3 degrees to fly the exact ground track.

Much of a landing pattern involves climbs and descents. You climb out along the *upwind* leg after takeoff; you descend along the *base* and *final* legs for the landing. Drift correction is critical. Unbridled drift during the climbout can blow your plane toward trees or other obstacles bordering a narrow airstrip, or drift you across the path of another departing aircraft (Fig. 2-6). If the plane drifts on base leg or final leg, the resulting erratic pattern disturbs your timing to the point that a good landing is improbable (Fig. 2-7).

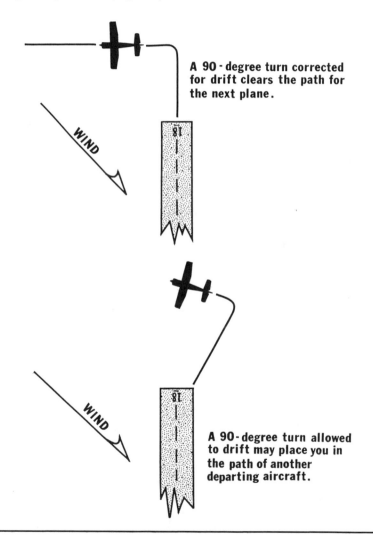

A 90-degree turn corrected for drift clears the path for the next plane.

WIND

A 90-degree turn allowed to drift may place you in the path of another departing aircraft.

WIND

Fig. 2-6. An unchecked crosswind could drift you into the depar-ture path of a following airplane.

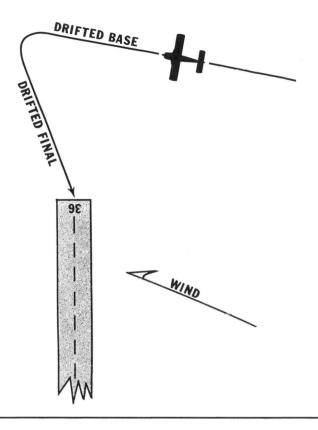

Fig. 2-7. Wind drift within the pattern will complicate your landing.

It is necessary, then, to control drift while climbing and descending. The flight-test examiner may ask to see you do it. Once alongside your same straight road, establish a crab angle in cruise flight. Next, enter a climb at best-rate-of-climb airspeed. When the plane slows to climb speed, naturally, increase the crab angle. Climb 500 feet as you fly along the road, correcting wind drift with a crab. Then, level off, do a turn-about, and correct in the other direction during a 500-foot descent (from 1500 AGL) at approach speed and power.

While most of your inflight tracking is by means of the crab, it is best to use the *slip* for the short-term wind correction during the descent and touchdown to the landing.

SLIPPING ALONG A ROAD

Once you master the slip, you will get no end of pleasure putting it into play. It's fun—killing the drift with a flick of aileron and holding the nose straight down the center line with a nudge of the rudder.

As stated earlier, the slip is used to correct for drift during the last moments of an approach to a crosswind landing. It is the only safe way to make a drift-free crosswind touchdown. The flight-test examiner may ask you to demonstrate a slip. The maneuver (1) effectively stops drift (a plane that touches down while drifting is very apt to swerve) and (2) keeps the fuselage aligned with the runway center strip (a plane that lands in a crab also tends to swerve).

Many pilots who have trouble making good crosswind landings fail, simply because they never really learned how to slip. Often, these pilots had to try and learn the slip *while* actually learning the crosswind landing. This is a bad learning situation at best—only a few seconds of exposure to a complex maneuver, low altitude, and anxiety about that slab of runway rushing toward you. If you really want to learn the slip, you need a better learning environment than that. Don't wait for the flight test.

Take your plane and the maneuver to a good working altitude— 1500 feet above the same road that you used for the crab. First, put your plane in slow flight (approach speed) alongside the road to keep it in sight, and kill drift with a crab. Second, apply carburetor heat and throttle back to 1700 rpm, or so, to start a 500-foot descent, holding drift initially with a crab.

Then, a shift from the crab to the slip (that's what happens on a short final). Move the plane into the slip by first banking the wings into the wind. In the next instant, press opposite rudder to keep the nose from turning with the bank. Now, use ailerons for drift and rudder for nose direction.

TRACKING THROUGH TURNS

When you track through turns rather than along straight lines you face two additional elements with which you must contend. First, if there is any wind present, you face a constantly changing *groundspeed* throughout the turn. And second, you must meet this groundspeed with a specific *rate of turn,* if you expect to fly a radius of turn that matches the desired ground track.

To better understand how groundspeed and rate of turn combine to produce a needed radius of turn, imagine yourself in the cockpit of a race car: The first lap is run and you are up to speed at 120 mph, approaching turn number one again.

Now, if that 180-degree turn is a quarter mile in length, your 2-mile-a-minute speed covers it in about 8 seconds. A rate of turn of about 22 degrees per second is called for. That's fast maneuvering. Of course, you don't take the time to calculate the needed rate to turn (and you don't do it in the plane either). But, you *do* realize that a rapid rate of turn is needed for the high speed. So, you enter the turn with a hefty twist of the steering wheel, which delivers the fast rate of turn.

Under the yellow caution flag you slow down and enter the turn at only 60 mph. The turn will now take 16 seconds, and you only need to nudge the steering wheel to produce the slower rate of turn that keeps you on the desired ground track.

This same relationship between groundspeed and rate of turn exists when you fly a plane through a turn when wind is present. In the plane, however, the groundspeed changes *while* you are turning, and you regulate the rate of turn with ailerons and rudder, rather than with a steering wheel.

Now, back to flying your plane: You are at 1000 feet, planning to fly a turn through a 180-degree ground track, with wind present (Fig. 2-8).

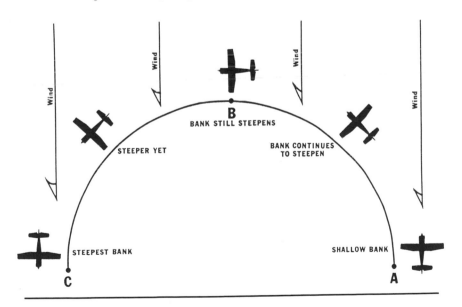

Fig. 2-8. When tracking through a turn, match your rate of turn with the groundspeed.

Match your rate of turn to the constantly changing groundspeed. As you enter the turn at point A, you face a direct headwind that produces the slowest groundspeed. You need only the shallowest bank for the slowest rate of turn. But as you continue into the turn, the headwind component lessens immediately, letting you pick up groundspeed. You need to immediately start to steepen the bank with coordinated rudder and ailerons.

The headwind component continues to diminish as you fly the arc toward point B, allowing the groundspeed to increase constantly. Continue to increase your bank with additional aileron and rudder pressures to keep increasing your rate of turn.

At point B you are in a medium-banked turn; the diminished headwind component shifts to that of a slight, initial tailwind.

Of course, as the wind shifts more and more to the tail, groundspeed increases, and the bank needs to continue to steepen. The steepest bank of the turn, naturally, occurs at point C, with a direct tailwind and the highest groundspeed.

Now, let's fly through some of the ground reference flight-test maneuvers.

■ TURNS AROUND A POINT

I think that *turns around a point* are the easiest of the ground reference maneuvers to fly. I say this because its ground track is the easiest to visualize—if you select good landmarks.

■ Landmarking the Maneuver

It makes little difference whether you are flying the traffic pattern at the airport, or turns around a point on the flight test. It is easier if you can visualize the ground track needed, and then choose the right landmarks with which to delineate the desired ground track.

Turns around a point involve flying a circle around a central landmark and correcting for wind drift so that a constant radius is maintained. If this circular course could be painted on the ground, most pilots would have little trouble tracking it, because they could *see* the path they needed to fly and would no longer be concerned with the wind direction or how and when to bank the wings. They would simply do whatever was necessary to stay on an easily recognized circular course.

It is important that you visualize the circle you want to fly during the flight test. Select and use landmarks that graphically mark the maneuver's ground track.

The best situation would be to use a road intersection as your centerpoint, with a landmark along each road for your four checkpoints. These four checkpoints—telephone poles; fence lines; buildings—must be equidistant (between 300 and 500 feet) from the centerpoint (Fig. 2-9). Then join the checkpoints together with an imaginary curved line for a prominent circular ground track. This circle is large enough to accommodate most light planes, yet is tight enough to require a reasonably steep bank in the turn.

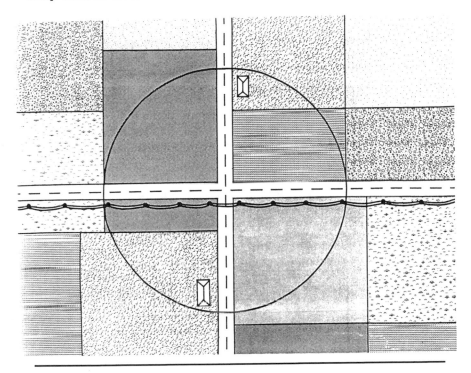

Fig. 2-9. Mark the imaginary circular path of the ground track for your turns around a point with good landmarks.

Clearly delineating the maneuver's ground track provides some fringe benefits. It is easy to loose track of the wind direction or become geographically disoriented after flying a turn or two around the centerpoint. The road intersection helps solve both problems. Before entering the maneuver, determine from your ground references just how the wind direction intersects the crossroad. Then, it is easier to keep tabs on the wind as you circle.

The crossroad will also help you stay oriented in the turn. Use the roads for directional guidance. You will then find it easy to enter the maneuver parallel to one of the roads, count the desired number of turns as you make them, and roll out on the entry heading.

■ Flying the Maneuver

When flying through turns around a point, keep in mind the four primary areas of concern facing the pilot:

1. Matching bank to groundspeed.
2. Correcting for wind drift.
3. Maintaining altitude.
4. Coordinating rudder and aileron movements.

Matching Bank to Groundspeed

When correcting for a varying groundspeed in this maneuver, you need to remember that *the slower the groundspeed, the shallower the bank; the faster the groundspeed, the steeper the bank.* For purposes of describing how the bank must vary to accommodate a changing ground-speed, start with the airplane at point 1 in Figure 2-10. (The banks shown are for illustration only. It would take a definite combination of wind, radius of turn, and speed for these exact banks to produce a constant-radius turn around a point.)

At point 1, the plane is in the direct headwind position and at its lowest groundspeed. Thus, it is at this point that the shallowest bank is required (20 degrees in the diagram). As the plane passes through point 1, it immediately starts to lose the direct headwind, and the groundspeed starts to increase. This also means that as the plane passes through point 1, the bank must begin to steepen to match the increasing groundspeed through points 2, 3, and 4 as the plane first loses the headwind and then finally turns toward the tailwind (the bank must increase as the ground-speed increases).

At point 5 the plane is momentarily in a direct tailwind and experiences the fastest groundspeed (the bank is 40 degrees, the steepest, at this point).

As the plane passes through point 5, it immediately starts to lose the direct tailwind, and the groundspeed starts to decrease. (This also means that as the plane passes through point 5, the bank must begin to shallow, if it is to match the decreasing groundspeed.) The groundspeed will continue to decrease through points 6, 7, and 8, as it loses the tailwind — it then starts to pick up a headwind. The bank must continue to shallow as the plane continues to lose groundspeed as it reaches the maneuver's point of entry.

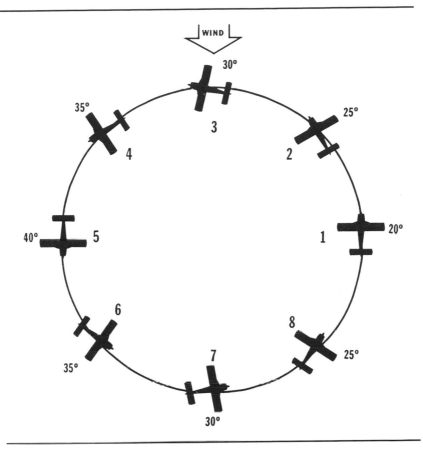

Fig. 2-10. Your bank must steepen or shallow to match changes in groundspeed.

Correcting for Wind Drift

Many pilots taking the flight test mistakenly feel that the steep bank should occur at point 7 and the shallow bank at point 3 in the illustration. The basis of this common misconception is the mistaken belief that the bank is varied to prevent drift. But, the bank is varied only to change the rate of turn needed to match groundspeed with the desired ground track. In turns around a point, you prevent drift by crabbing just as you did in tracking along a straight road.

In a turn, the crab just *looks* different. Notice how the plane's nose angles toward the wind when flying around a point (Fig. 2-11). The pilot steers the plane so that the *apparent* path will follow the dotted lines, just as when tracking a road. But whether in a turn or along a straight

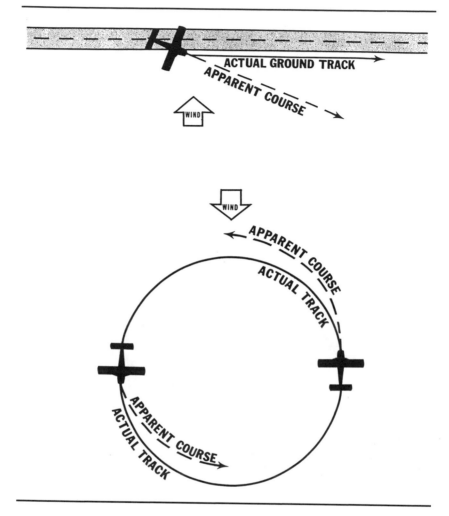

Fig. 2-11. A wind correction angle prevents drift in the turn, just as it does when tracking a straight line.

line, a crab is a crab, and it accomplishes the same end—it prevents drift.

Maintaining Altitude

Some pilots taking the test let the plane's altitude vary while flying turns around a point. This normally happens because they either fail to change elevator pressure as they steepen and shallow the turn, or they neglect their reference to the horizon as they circle the centerpoint.

You need to apply additional stick pressure during those moments of increasing bank. If you do not, the plane will lose altitude. Conversely, relax stick pressure as you shallow the bank, or you will climb. Remember that both stick pressure and bank are in a constant state of change when flying the maneuver.

A common error during turns is to think in terms of shallow bank and steep bank with *no* bank change in between. The bank, however, must constantly change, simply because the relative wind direction and groundspeed constantly change (this means a constantly moving stick).

Use your outside horizon reference to maintain pitch control, as you do in normal turns at altitude. Learn to divide your attention between the ground reference points and the horizon—this takes a conscious effort on your part.

Coordinating Controls

A few pilots seem to forget everything they learned about coordinating ailerons and rudder, once they enter turns around a point. Don't let this happen to you, because skids and slips will usually develop for two reasons. First, the inboard wingtip covers the centerpoint during part of the 360-degree turn around the point. This results in you trying to move that wingtip with aileron for a "look-see" and you then force the turn with rudder. Second, you get so involved with when and where to move the controls that you forget the actual ground track you are trying to follow. Both of these uncoordinated controls movements are caused by the pilot not adequately visualizing the ground track.

The selection of four suitable checkpoints equidistant from the centerpoint is essential, because the wingtip *will* cover the centerpoint during a portion of the turn. (During a *right* turn around a point, the centerpoint and at least one checkpoint is hidden during *most* of the turn. Guess which way the examiner will have you fly the maneuver on the flight test.)

■ A Few Tips

Turns around a point are considered by many to be one of the most difficult of the flight-test maneuvers. Personally, I don't think this is the case if you follow a few tips:

1. Take the time to find a good centerpoint with adequate checkpoints. Once you visualize the circle on the ground, the maneuver really becomes quite easy. Become adept at finding good checkpoints as you become aware of the subtle pattern on the ground. (The examiner may not let you fly to your favorite spot.)

2. Refrain from flying the maneuver as numerous consecutive 360-degree turns. Two, or at the most three, turns around a point is enough. Prolonging the maneuver longer may prove uncomfortable.

3. Use an altitude of 1000 feet AGL. This is high enough for safety and relaxation, yet low enough to let you easily detect any drift.

4. *Know* the wind direction before you enter the maneuver.

Safety First

When making turns around a point you fly relatively close to the ground while much of your attention is directed toward the terrain, away from the plane. Stay relaxed during your demonstration, and put four defensive measures into play. First, scout the area for tall obstructions or ridges. Second, before you start the maneuver, scan the practice area for other low-altitude aircraft that may be practicing *their* ground reference maneuvers. Third, keep a "corner of your eye" on the altimeter—if you lose as much as 200 feet, don't try to save the maneuver; you are too close to the ground. Simply break off the maneuver, regain your altitude, and reenter for another try. Your examiner will likely allow this without discredit. And finally, avoid bank angles of 45 degrees or greater. This excessive attitude, with your attentions directed toward the ground, can result in a *rapid* loss of altitude.

Once you have successfully flown turns around a point, you have demonstrated far more than the ability to correct for wind drift. You have also displayed competence in two important areas of flying. First, you have shown the ability to maneuver your plane by "feel" while your attention is elsewhere. An example of this could occur when low on base leg, you suddenly see a converging aircraft coming in on final—and closing fast. Now, close to the ground, you must quickly select a ground track that offers safety, then turn your plane away from danger *while* you keep an eye on that other aircraft.

The second area of competence is the ability to distinguish subtle landmarks and make use of them. It is your first step of immediate importance toward navigation, and the ability to use subtle terrain features makes life so much easier as you move into the next ground reference maneuver of the flight test, S-turns across a road.

S-TURNS ACROSS A ROAD

Simply put, it is fun to fly S-turns across a road. The maneuver offers fast, close-packed action, which is not often found in most training exercises—"blasting" across the road, bending into the first turn, steepening further to match the mounting groundspeed, keeping the radius smooth, cutting across the road once more, and then throwing the bank to the other side to keep the S-turn going in the next direction.

S-turns match the discipline and finesse of turns around a point with the dash and razzle-dazzle of aviation's golden age of pylon racing. It is impossible to pull through the maneuver's turns and not identify with the pylon racers and pilots such as Roscoe Turner, Amelia Erhart, and Jimmy Doolittle.

■ Flying the Maneuver

S-turns across a road is a precision maneuver in every sense of the term. This maneuver is not as spectacular as airshow maneuvers, true. But the mastery of S-turns demands the same measure of effort and discipline as do the airshow stunts.

S-turns (Fig. 2-12) are a series of alternating 180-degree turns across a road (or similar long straight line). To qualify as a perfectly executed flight-test maneuver, S-turns demand four essential factors of precision flying:

1. Each semicircle must be a smooth curve of equal radius.
2. Altitude must remain constant throughout the turns and at the crossovers.
3. The turns must be timed so that the wings align with the road at each crossover. (And if the wings do *not* align, you should break away and re-enter, for the succeeding turn will not work.)
4. The controls must remain coordinated throughout the maneuver.

Probably the best way to explain how to fly the S-turn is by means of a simple line drawing (Fig. 2-13) and an explanation of the steps to make throughout the maneuver.

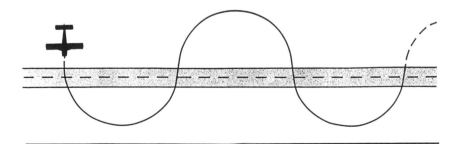

Fig. 2-12. S-turns are a series of 180-degree turns across a road.

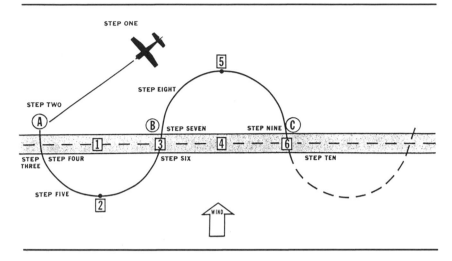

Fig. 2-13. **Plan to fly the S-turn as a series of steps.**

Step One. Select a long road as near square with the wind as possible. Clear the area for obstacles and other low-altitude aircraft and descend to a maneuvering altitude of 1000 feet AGL.

Step Two. Approach point A, into the wind at 1000 feet AGL, with the wings parallel with the road.

Step Three. As you cross the road, quickly pick landmarks 1, 2, and 3 to mark the centerpoint of the turn, the radius of the turn, and the crossover point B.

Step Four. Begin turning into the first loop of your S as the road passes under your seat. In our example, you are entering into the wind. Of course, it makes no difference whether you start against or with the wind — you should feel comfortable with either entry.

But since you are now entering upwind, be aware of a relatively low initial groundspeed. This calls for a relatively shallow bank at the outset, accompanied by only slight back pressure to keep the altitude consistent.

Step Five. Vary aileron, rudder, and elevator pressure to maintain the desired ground track around landmark 2. Since you immediately start to reduce the headwind component, your groundspeed steadily increases, and you, therefore, must steadily increase bank with aileron and rudder if you are to stay on track (as you learned in turns around a point). You must also continue to apply increasing back pressure on the stick to maintain altitude. (As you bank steeper and steeper, the plane's weight steadily increases in aerodynamic G loading. This factor must be met with an increasing angle of attack.)

Two items of caution should be mentioned here. First, remember, you are maneuvering fairly close to the ground, so if your altitude varies by as much as 200 feet, just admit that you blew it, break off, and set up for another entry. Your examiner will probably not fault you. Second, if it becomes apparent that you need to increase the bank more than 45 degrees to meet point B, break away and try again. Again, the examiner may see this move as evidence of good judgement.

Step Six. Take that instant before you reach the crossover point B to select landmarks 4, 5, and 6 to delineate the midpoint of the second half of the S, the ground track, and the crossover point.

Step Seven. The crossover point B needs careful foot- and hand-work, for it is here that your control coordination is put to the test — not only are you making a coordinated roll-out from a left turn, but you must coordinate the controls immediately and move into a right turn. During this roll-out and roll-in, you must coordinate elevator action with the changing forces so that you maintain altitude.

The stick pressures that deliver this needed elevator action are, themselves, ever changing. As you begin to roll out of your steep turn, aerodynamic loading lessens. During the roll-out, you need to reduce stick pressure at a constant rate, which has the elevator neutral just as the wings level at the crossover point.

When the road passes beneath your seat at the crossover point you again need to start applying stick pressure as you roll into the right turn.

Step Eight. Vary aileron, rudder, and stick pressure to maintain the ground track and altitude toward landmark 5. In our example, you will immediately start to loose the tailwind component and groundspeed slows. To keep the ground track, you must start a steady reduction in bank, with a diminishing stick pressure. Bank and stick pressure will continue to turn toward the headwind at point B. Aileron, rudder, and stick pressures are ever-changing throughout the S-turns across a road; there is never a moment when they are fixed.

Step Nine. Take that instant before you complete your first S, at point C, to select landmarks that delineate the succeeding portion of the next S-turn. Clear for traffic.

Step Ten. Continue a series of several S-turns along your road. Keep in mind three safety measures: (1) break away anytime you allow the altitude to stray 200 feet or more, (2) break off the maneuver anytime it appears that a bank in excess of 45 degrees is needed to maintain the ground track, and (3) clear for traffic at each crossover point.

■ Pilot Skills Developed

The hours you spend aloft perfecting S-turns, before and after solo and while preparing for the flight test, develop four profound piloting skills that will serve you well throughout your pilot career: (1) the ability to quickly pick up and use subtle landmarks, (2) a keen sense of timing, (3) a remarkable sense for instinctive control coordination, and (4) the flyer's sixth-sense, the "feel for flying."

Let's look at each of these skills in turn as they apply to the maneuver.

Landmark Recognition

As in any ground reference maneuver, the ground track, to be flown precisely, must first be visualized. You must develop the ability to select and use landmarks to delineate the course. But in S-turns, there is a significant difference from turns around a point. In performing S-turns, you must find and establish your landmarks *quickly*—very quickly—*while* you are flying the maneuver.

Figure 2-14 shows the relationship of the ground track to the landmarks. As you approach the road to enter the maneuver at point A, you have only the briefest moment of level-wing flight to pick out three landmarks. First, you need to select landmark 1 to establish the centerpoint of the turn. Second, in the "blink of an eye," you must find land-

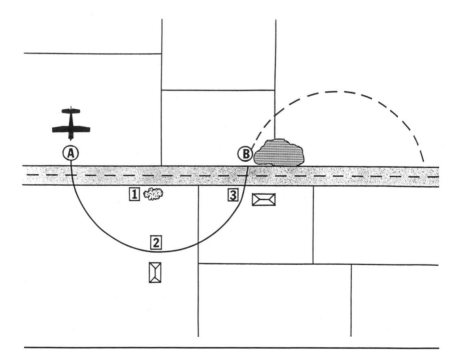

Fig. 2-14. Landmarking the maneuver helps visualize its ground track.

mark 2, which helps delineate the radius of the turn. Then you must quickly pick landmark 3, the crossover at point B, halfway through your S-turn.

Upon reaching point B, of course, you repeat the process of picking the landmarks to complete the first S-turn. The selection process is then repeated over and over as you continue "S-ing" down your mile or so of straight road.

You see the problems, don't you. At each crossover you have just an instant of level-wing viewing to pick your marks. You have to do it quickly and correctly if your S-turns are to work. And, at each crossover, you must coordinate the controls during a rapid complete change of direction, while you strive to maintain an exact altitude and clear for traffic. It isn't easy, and there are no shortcuts to make it easy. It's up to you and hard work. One helpful hint is, don't be too picky about the landmarks you select. Chances are, strong landmarks will not be available when and where you need them. Be content with subtle marks on the terrain — a patch with a different texture, a ditch, or even something you can only half imagine.

Learning to detect and use subtle landmarks takes practice. But that's a part of what the maneuver is all about. And the skill will serve you well throughout your flying. An example of how it can be used is if you are entering a landing pattern on a gusty day that tries to bounce you into the next county, you can visualize the pattern's ground track and fly it accurately under trying circumstances.

Do all pilots have the ability to detect and use subtle landmarks? No, they do not. But, the ability to do so is one very noticeable hallmark that separates the superb flyer from the average. And my sincere wish is that you become a superb pilot.

Sense of Timing

Once you have visualized the perfect semicircles of your S, you must then fly that precise ground track. Your success in this effort depends, to a large extent, on a keen sense of timing. Take another look at the ground track relative to the wind, and the problem of timing is obvious (Fig. 2-15).

Turns around a point taught you about groundspeed versus angle of bank; each loop of the S involves a constant change of bank, with a constantly changing rate of turn. The problem, of course, is to manage this change so that each crossover is achieved with the wings parallel to the road *and* each loop is a symmetrical semicircle. If the rate of turn is mismanaged, the plane will not meet the crossover point with wings parallel to the road. For example, if the pilot either delays the initial rate

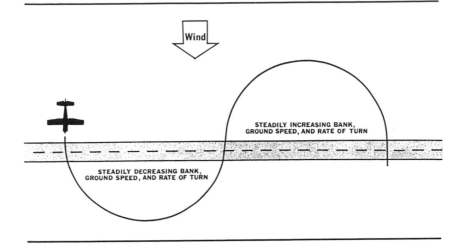

Fig. 2-15. Vary the bank to match the plane's rate of turn with the changing groundspeed.

of roll-in or rushes the roll-out, the turn will not be completed in time— the wings will not be parallel to the road at the crossover point but will angle away from the turn (Fig. 2-16). But if, on the other hand, the pilot either rushes the roll-in or delays the roll-out, the turn is finished before the pilot is ready. At the crossover point the wings turn into the turn (Fig. 2-17).

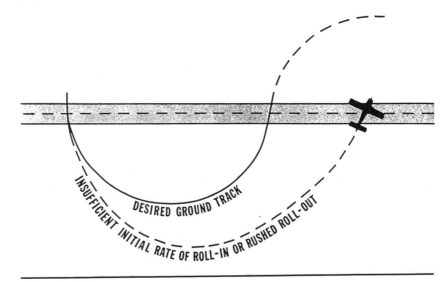

Fig. 2-16. Delaying the roll-in or rushing the roll-out tracks the plane to the outside of the curve and spoils the symmetry of the S-turn.

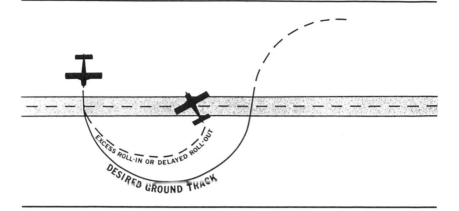

Fig. 2-17. A rushed roll-in or delayed roll-out tracks the plane to the inside of the curve and spoils the symmetry of the S-turn.

Timing is all important, and to facilitate this timing in flying S-turns, the pilot taking the flight test must remain aware of two factors within the maneuver. First, you must constantly be aware of the wind relative to the ground track of the next moment, and adjust your rate of turn instantly to the changing groundspeed.

Second, you must keep the ground track leading to the visualized crossover point. Then it is easy to keep changing the bank and rate of turn to fly that visualized line. A common mistake that is made here is that a student often tries to make the first three-fourths of the loop with a fixed, medium-banked turn. Then, in the closing seconds before the crossover point arrives, they do whatever is necessary with the bank to get the plane square with the road at the crossover point. This invariably produces a non-symmetrical semicircle. The loop ends up either flattened (on the windward side of the road) or sharpened like the point of an egg (on the lee side of the road). The flight examiner will not appreciate either ground track.

Control Coordination

In addition to timing, S-turns require *control coordination.* During this maneuver, the ailerons, rudder, and elevator are constantly changing to meet the requirements of an ever-changing groundspeed. In each semicircle of the S, the bank changes from shallow to medium to steep as the wind direction shifts from a nosewind to a crosswind to a tailwind. If you are to hold altitude, stick pressure must change simultaneously as the bank changes.

Turns around a point presented this same challenge but S-turns add a degree of difficulty—keeping ailerons, elevator, and rudder coordinated through alternating, complete, rapid reversals of direction. There is no doubt that you will eventually develop an instinctive "feel" for the controls in your flying.

The Feel for Flying

The S-turns provide an excellent chance for you to develop your "feel for flying," which is something most pilots want to see in their own flying. Yet, many find it difficult to develop. Those pilots who experience difficulty usually do so, simply because they do not know exactly what to look for—they may even think that a feel for flying is only "theory." They don't understand what it means to have a "feel" for flying, and until they do understand, the "feel" will be elusive.

So, here are four specific factors that constitute this "feel for flying," so you know exactly what to look for.

1. You are aware of the environment surrounding you and your plane—the sights, sounds, and sensations that inform.

2. You are aware of the aerodynamic forces that affect your plane.

3. You *anticipate* the effects that the environment and aerodynamic forces will have on the plane, before those effects occur.

4. You act to compensate for these effects the instant they occur.

The feel for flying, then, is a total awareness of conditions and forces and your ability to compensate before they adversely affect your plane. It is the simple difference between *managing* a situation and *correcting* a situation already gone awry.

Remember, the feel for flying is a learnable skill. And, you don't need to wait until you are a 10,000-hour "pelican" either. Once you know the four factors to look for, then make a conscious effort to develop this feel, early in your training.

Performing S-turns holds a lifetime of enjoyment. There will be days—long after the flight test—when you need to relax. When you are aloft and a long straight road presents itself, settle into your seat, hang on to the stick and throttle, curl your toes around the rudder pedals, and take your plane into the S-turns.

RECTANGULAR COURSE

If S-turns are the razzle-dazzle of flying, then the rectangular course is staid by comparison like being seated comfortably behind the control wheel of a lumbering DC-3. The maneuver is easy, relaxing, and seems to play out in slow motion (plan to take an apple along).

■ The Ground Track

The rectangular course simulates the airport traffic pattern (Fig. 2-18). It only requires corrections for wind and some turns while you maintain the ground track.

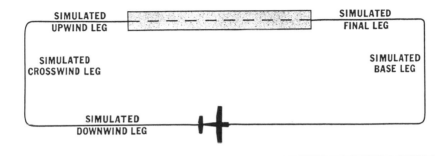

Fig. 2-18. The rectangular course simulates the airport traffic pattern.

■ Selecting a Site

As when selecting *any* site for ground reference maneuvers, scout the area for obstacles or populated areas that might be disturbed by a low-flying airplane. Try to pick a pattern of straight lines or roads that provide a rectangle of about a mile on the long side and a half mile on the short side. Don't worry too much about having your pattern square with the wind; it rarely happens, anyhow, in a traffic pattern. And by this time, you are skilled enough with the wind that you won't mind what nature hands you.

Keep two precautionary measures in mind. First, don't forget to use those 90-degree turns for their intended purpose—to clear for traffic. There is no better time for a mid-air collision than those moments during a ground reference maneuver when much of your attention is directed toward the terrain. And we all know too well how much a mid-air can spoil your whole afternoon.

The second precaution concerns altitude. Plan your altitude no lower than 1000 feet AGL.

■ Flying the Maneuver

Fly the rectangular course in reference to 12 keypoints (Fig. 2-19).

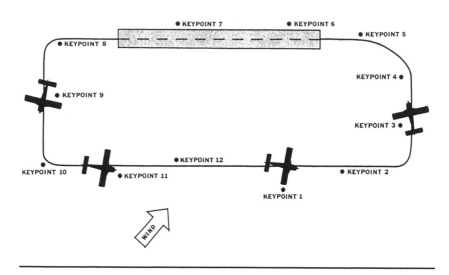

Fig. 2-19. Fly the rectangular course against 12 key points.

Keypoint 1. At about 1000 feet AGL establish a wind correction (crab) angle that holds the ground track for the downwind leg at cruise speed. In Figure 2-19 there is a quartering right crosswind, therefore, you would crab to the right. If that wind velocity was 15 knots, your correction angle would be about 5 degrees.

Keypoint 2. As you near the completion of the downwind leg, begin turning to the simulated base leg, keeping in mind the wind direction and the expected drift.

But, here you handle drift differently than you did in S-turns. In flying the S-turns you handled the changing groundspeed with a change in bank — meeting the slower groundspeed with a shallow bank and the faster groundspeed with a steeper bank. In the rectangular course, however, because you are simulating the airport traffic pattern fairly close to the ground, you want to avoid steep turns. Therefore, you need a method to handle the drift in those pattern turns, in which there is no need for steep turns. There is a method at hand.

Begin your turn to base earlier than a symmetrical turn would call for. (This turn will *not* be symmetrical, but it does the job.) And, even though you are flying with a tailwind, just roll into a modest bank and *let* the wind drift you toward the base leg position. Then, halfway through the turn, start rolling out as you see the base leg ground track aligning with your plane. It is quite simple to do — use your sense of timing that you demonstrated in the S-turns.

Keypoint 3. Roll out on base leg and establish a crab angle to kill the drift. The simplest way to establish this angle is to let your turn carry you a few degrees *past* the alignment with base leg.

Keypoint 4. Start your turn from the base to final. Handle wind drift in the turn with no more than a modest bank. In our example (with a tailwind), start the turn early and let the wind drift you toward the final leg. (If you have a headwind on base, handle the drift in the customary manner—with a shallow turn to final.)

Keypoint 5. Roll out on the simulated final. Again, in our example, roll out a few degrees *past* final leg alignment to establish the needed crab angle.

Keypoint 6. Fly down your simulated runway at 1000 feet AGL, using a wind correction angle to stay aligned on your simulated center strip.

Keypoint 7. Make any small adjustments to your wind correction angle, needed to maintain your upwind track.

Keypoint 8. Turn from upwind to the crosswind leg. Since you are turning into the wind, a relatively shallow bank should be sufficient. Roll out a few degrees shy of the crosswind track and crab just enough to kill drift.

Keypoint 9. On the crosswind leg, correct for wind drift with a crab. Make any small adjustments to your wind correction angle as needed to maintain the crosswind track.

Keypoint 10. Roll out of the turn with an established crab angle.

Keypoint 11. Make small adjustments to your wind correction angle if further correction is needed to maintain the downwind track.

Keypoint 12. Complete the maneuver on track, within 100 feet of entry altitude, and within 10 knots of entry airspeed. Remember to clear for traffic with each 90-degree change of direction throughout the maneuver.

IN REVIEW

■ Tracking Along Lines

- Select a road as near square with the wind as possible.
- Evaluate wind direction and velocity from ground references.
- At 1000 feet, the surface wind will shift about 20 degrees to the right and increase in velocity about 50%.
- Plan a crab angle of about 1 degree for each 2 knots of crosswind component.
- The *crab* angle must *increase* with any *decrease* in airspeed.

The *slip* is designed to prevent drift while maintaining a straight-ahead heading.

Ailerons in the slip prevent drift.

Rudder in the slip steers the nose straight.

■ Common mistake: Using too little opposite rudder for the aileron applied.

Airspeed tends to slow down in a slip.

Tracking Through Turns

When tracking through a turn, wind constantly changes your ground-speed.

You must match each change of groundspeed with a specific rate of turn.

Steepening the bank increases the rate of turn, shallowing the bank decreases the rate of turn.

Turns Around a Point

It is easier to track through a turn around a point if you select landmarks to *visualize* the ground track.

Use crossroads and landmarks to delineate the path of turns around a point.

When correcting for a varying groundspeed through a turn, remember that the slower the groundspeed, the shallower the bank and the faster the groundspeed, the steeper the bank.

In turns around a point, the pilot prevents drift by crabbing, just as in tracking along a road.

To maintain altitude around the point, increase stick pressure as you steepen the bank or decrease stick pressure as you lessen the bank.

■ Fly turns around a point no lower than 1000 feet AGL.

Avoid angles of bank in excess of 45 degrees.

S-Turns Across a Road

■ Each semicircle of the S-turn must be smooth and of equal radius.

Altitude must remain constant throughout the turns and at each crossover point.

■ Turns must be timed so that the wings align with the road at each crossover point.

■ Controls must remain coordinated throughout the maneuver.

■ S-turns develop the skill to quickly see and use subtle landmarks.

Avoid altitudes below 1000 feet AGL, and banks in excess of 45 degrees.

The Rectangular Course

The rectangular course simulates (at a safe altitude) the airport traffic pattern.

Choose roads about 1 mile in length for the long side of the rectangular pattern, with short sides of a half mile to delineate your pattern.

Plan your entry altitude so that the maneuver is no lower than 1000 feet AGL.

Plan to handle wind correction in the turns without using steep turns. Learn to let the wind drift you into position.

■ Plan to roll out of each turn with the crab angle for the new leg already established.

FLIGHT-TEST GUIDELINES

Turns Around a Point

Turns around a point are designed to evaluate three important pilot skills:

1. The ability to select and use landmarks that assure the desired ground track.
2. The skill that allows proper control coordination and altitude management while your attention is directed away from the cockpit.
3. The ability to correct for wind drift while contending with an ever-changing groundspeed.

Demonstrate the maneuver to your flight examiner in a sequence of five steps:

1. While at cruise altitude, select a pattern of landmarks that delineate the appropriate ground track. Select a pattern that provides a center-point and four equidistant points that mark your desired radius (see Fig. 2-9). Determine the wind's direction across the pattern.
2. Before descending toward your selected pattern, search the area for low-flying aircraft and tall obstacles.
3. Descend to a working altitude—1000 feet AGL is appropriate. This altitude is low enough to let you immediately detect wind drift, yet

high enough to provide safety and prevent disturbance to the residents below.

4. Enter the maneuver's ground track over one of the four delineating landmarks.
5. Fly the maneuver. Manage the changing groundspeeds, wind drift, and pitch control (see Fig. 2-10).

Flight-Test Tolerances

1. Maintain coordinated controls throughout the maneuver.
2. Select suitable ground reference points.
3. Be able to divide attention between aircraft control and the ground track.
4. Hold entry altitude within 100 feet throughout the maneuver.
5. Hold entry airspeed within 10 knots throughout the maneuver.
6. Be able to explain the procedures used to compensate for wind drift in turns around a point.

■ S-Turns Across a Road

As in the previous maneuver, S-turns across a road test a pilot's skill to make good use of landmarks, correct for wind drift, and to fly accurately with attention divided between the plane and terrain. In addition, this maneuver allows the pilot to demonstrate a keen sense of timing — accurate timing keeps the wings parallel to the road at each crossover point (see Fig. 2-12).

Plan your flight-test demonstration as a sequence of five steps:

1. While at cruise altitude, select a straight stretch of road 1 mile or so in length. Select landmarks that provide a center crossover point for the first full S-turn, with supporting marks that delineate your desired radius of turn. On completing the first full pattern, *quickly* select similar landmarks for the subsequent S-turn. Determine the wind's direction across the pattern.
2. Before descending toward your maneuvering altitude, search the area for tall obstructions or low-flying planes.
3. Descend to your maneuvering altitude — 1000 feet AGL is appropriate for most light training aircraft.
4. Enter the ground track over the road above the first landmark, which delineates your desired radius of turn — the wings must parallel the road at this entry point.
5. Fly the maneuver. Manage the changing groundspeeds (see Fig. 2-15).

Flight-Test Tolerances

1. Enter the maneuver with wings parallel to the road.
2. Apply necessary wind drift correction to produce symmetrical S-turns. Reverse the direction of the turn directly over the road.
3. Maintain altitude within 100 feet of the entry height.
4. Maintain airspeed within 10 knots of entry speed.
5. Maintain coordinated controls throughout the full S-turn.
6. Be able to explain the wind drift corrections needed.

■ Rectangular Course

Flying the rectangular course simulates the airport traffic pattern. This maneuver tests your ability to clearly visualize the ground track of each leg and to effectively compensate for wind to maintain the ground track.

Fly your test demonstration as a series of four steps:

1. While at cruise altitude, select a rectangular ground pattern approximately 1 mile in length and a half mile in width. Determine the wind direction across your pattern.
2. Clear the area for low-flying aircraft and high obstacles.
3. Descend to 1000 feet AGL and enter one of the longer legs at a 45-degree angle.
4. Fly the pattern, correcting for wind drift as indicated in Figure 2-19.

Flight-Test Tolerances

1. Be able to correct for wind, to maintain straight pattern legs.
2. Maintain altitude within 100 feet of the entry height.
3. Maintain airspeed within 10 knots of the entry speed.
4. Maintain all bank angles within 45 degrees.
5. Maintain coordinated controls throughout the pattern.
6. Be able to explain wind drift corrections needed.

3.

Takeoffs and Climbs

The flight examiner is very interested in seeing precision takeoff procedures and actions, because the FAA has pinpointed the takeoff as the
time and place for the majority of flying accidents. The reasons are
many: the plane is *close* to the ground at slow flying speed; cockpit
activity is pressed for time; the airplane must perform at maximum
capability; nearby obstacles often reach into the low departure altitude;
many others. The flight examiner never wants to see an applicant fall
prey to the pitfalls of this very critical maneuver and, therefore, pays
close attention to the expertise and precision given to takeoff procedures
and actions.

A word of caution. Takeoffs are discussed in a single chapter as they
are performed on the flight test. But you must remember that the demonstrations are normally conducted under ideal conditions, with difficulties only simulated. For example, short-field and soft-field procedures
are normally demonstrated from long, paved runways with only an
imaginary obstacle reaching upward toward the departure path. Real
difficulties such as downwind takeoff, a strong crosswind, a high-elevation departure, a rain-slick runway, or wake turbulence are seldom encountered on a flight-test demonstration. These, along with many other
real-life difficulties are the elements that stand between *flight-test demonstrations* and *operational reality.* All these elements affecting takeoff
procedures *cannot* be covered in a single chapter. It takes a book. And I

have written such a book: *Making Perfect Takeoffs in Light Airplanes* (Iowa State University Press). I urge you to study this text. And again, remember that the discussion of takeoffs herein is fairly well confined to the demonstration of the departure procedures for the flight test.

NORMAL TAKEOFF AND CLIMBOUT

In discussing normal takeoff procedures for the flight test, I presume that you are in the run-up area with the pretakeoff checks completed and ready for "cleared for takeoff" clearance. Let's begin our demonstration at that point, and perform the maneuver as a sequence of seven specific steps:

1. Taking the runway.
2. Making the takeoff run.
3. Lifting off.
4. Flying the initial climbout.
5. Leaving the pattern.
6. Climbing to cruise altitude.
7. Leveling to cruise altitude and airspeed.

■ Taking the Runway

Perform this first step with precision. At a controlled airport (those with control towers), be ready to move onto the runway before you request takeoff clearance. The controller's "cleared for takeoff" implies a need for prompt action, for the tower may well have another plane on approach for landing. Any delayed cockpit activities such as closing a door or setting a navaid can easily interrupt the traffic flow. (The most common interruption that examiners mention is the time taken to hang the microphone on its panel clip. Many applicants spend precious moments on this unimportant task. Make *your* examiner smile — just drop the thing in your lap and get on with the business at hand.)

If you are starting your flight test from an uncontrolled field without the protection of a control tower, you must clear for traffic yourself. (Even when departing under the protection of a control tower, take advantage of the turns needed to position your plane on the runway. Look for traffic. Controllers can slip up.) The only way to do this is to taxi a full circle before you take the runway (along with a unicom announcement of your intentions). Don't be content to clear only the final and base legs of the runway that you intend to use. *All* runways are legally active at uncontrolled airports, even the *opposite* end of *your* runway. If other departing airplanes crowd the runway area, you may need to back-taxi down the taxiway for circling room.

Once you have cleared for traffic and are ready to take the runway, pick up the taxiway's yellow center line with your nosewheel. This taxi line will lead you directly to the white center stripes of the runway. The examiner wants to see you position your plane directly astride those stripes. This precise positioning is important for two reasons. First, runways are laid with a high crown for quick drainage. If you attempt to take off on the sloping side surface, you invite loss of directional control on the takeoff run. Second, when you begin your takeoff roll from dead center, you conserve runway width, which just may be needed in the event of an unexpected swerve during the run.

Once astride the center stripe with the nosewheel pointed straight ahead, bring the plane to a momentary stop. This gives you a short pause with which to gather thoughts toward the next step — making the takeoff run.

■ Making the Takeoff Run

The flight examiner is looking for four specific qualities in your takeoff run:

1. The most rapid acceleration to rotate speed.
2. Verification of normal engine performance.
3. Maintenance of positive directional control.
4. Correction for any crosswind.

Acceleration

A smooth hand on the throttle is your first step toward delivering a rapid acceleration. A throttle that is abruptly shoved in is very apt to feed fuel to the engine faster than the cylinders can digest it, and the engine will likely cough and spit, which significantly reduces acceleration. The best application of power is to smoothly advance the throttle as fast as the engine can accept it without giving any audible sign of misfiring.

Unnecessary deflection of the flight-control surfaces (aileron, rudder, elevator) adds significant drag, which retards acceleration. Allow the control surfaces to streamline with the wind unless there is a need for deflection that takes precedence over acceleration, such as the need for crosswind correction.

Some pilots develop the habit of "walking" the rudder to and fro for directional control during the takeoff run. This is not only an ineffective way to keep the plane rolling straight ahead, but each wiggle of that rudder adds drag, which retards acceleration. Other pilots attempt to use differential braking to keep the plane running straight. Don't make this mistake. The action not only invites a potentially dangerous swerve, but it also *drastically* retards acceleration.

Some pilots feel they can encourage acceleration by lifting the nose-wheel as soon as groundspeed permits — a mistake. Their erroneous thinking is to reduce the rolling friction of the nosewheel against the runway surface. When you lift the nosewheel, however, you also increase the wing's angle of attack to produce lift. A by-product of this lift is aerodynamic drag, which in turn retards acceleration early in the takeoff run. Conversely, however, do not press the control yoke forward. This throws nosewheel pressure against the runway and creates undue rolling friction, which adversely effects acceleration. It is best to let the control surfaces streamline unless there is a real need for control deflection, which takes precedence over acceleration.

Verification of Engine Performance

You should verify engine performance early in your takeoff run. Once full throttle is applied, glance quickly at the engine instruments. Does an anemic tachometer or abnormal oil temperature or pressure suggest an abort?

Maintenance of Directional Control

The flight examiner wants to see near-perfect directional control during your takeoff run for three reasons. First, the examiner knows that the best acceleration is gained from a straight and true takeoff run. Any control deflections needed to correct a swerve add drag. Second, the examiner wants to see runway width conserved. A lift-off from a swerve near the edge of the runway can aim the plane directly toward a nearby obstacle (parked aircraft, wind sock, hangar). And finally, the examiner knows that a relatively inexperienced pilot often cannot prevent a swerve, once started, from growing in magnitude. An airplane, unlike an automobile, does not have good road-handling qualities — only three points of surface contact, along with undersized tires, prevent that. It is quire possible for a swerving plane to dip a wingtip into the ground, and although a tattered wingtip rarely leads to personal injury, the cost of repair can dampen a pilot's enthusiasm for flying for months.

There is a fourth reason for preventing a swerve — one personal to you. You will soon be carrying passengers. Imagine the impact that a healthy swerve has on a passenger, even if no damage occurs. Imagine your passenger's perception—the sight of a runway suddenly gone askew, the sound of squealing tires, the apparently frantic action of the pilot trying to regain control. It's enough to make the hardiest passenger holler, "Lemme outa here!" And you haven't even gotten them into the *air,* yet. One thing is certain—passengers are most unforgiving. That passenger will never again want you as their pilot.

Now, many fledgling pilots try to maintain directional control by using the differential braking system. As previously stated, this is an unacceptable procedure. The rudder and the steerable nosewheel are the principle tools for directional control during the takeoff run. The simplest way to keep the brakes out of this steering action is to keep your heels on the floor. This prevents your instep from riding on the rudder with your toes touching the brakes. However, you can quickly slide your foot up, should brakes be needed to stop the plane.

During the takeoff run the two aerodynamic forces that will disturb direction control unless you take corrective action are called *slipstream effect* and *aileron drag.* Your examiner is most interested that you understand these forces and that you take measures to counter them.

SLIPSTREAM EFFECT.   The aerodynamic force of slipstream effect is created by the plane's propwash. The propeller blade that swings downward in the right-hand half of the propeller arc (as viewed from the cockpit) pushes the air under the fuselage. This slipstream of propwash then corkscrews up around the left side of the plane and flows upward to strike the left side of the vertical stabilizer (Fig. 3-1). This, naturally, tends to slew the plane to the left. Slipstream effect is most apparent during the initial moments of the run, right after full throttle is applied and ground travel is still slow. During these moments it takes quick

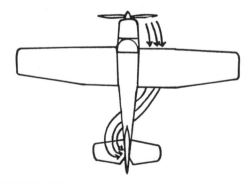

Fig. 3-1. Slipstream effect is most noticeable during the first few moments of the ground roll before takeoff.

right-rudder pressure to keep the plane directly astride the runway's center line.

Use just enough corrective right rudder to keep the plane running straight down the center line. And let me say here that the prime purpose of the center line is to provide accurate guidance to the pilot during the takeoffs and landings. Be aware of the line's purpose and use it to your best advantage.

AILERON DRAG. The flight examiner wants to see an awareness and response to the force that can disturb directional control on the takeoff run—aileron drag. Some pilots experience difficulty with directional control simply because they are unaware of this force and its effect.

Aileron drag (sometimes called *adverse yaw*) occurs when you turn the control wheel. It's effect is that when you turn the control wheel in one direction, the *airplane* turns in the opposite direction. If you turn the wheel to the right, for example, the right aileron deflects upward and is shielded from the onrushing air by the wing's curved upper surface (Fig. 3-2). The left aileron, however, deflects downward to dip beneath the wing's lower flat surface. This left aileron then digs into the onrushing air to turn the plane to the left.

Fig. 3-2. Here, the left aileron dips downward and protrudes into the slipstream.

The most common problem arises when you try to correct a swerve by instinctively turning the control wheel away from the direction of swerve. But this erroneous action only serves to intensify the swerve. To understand what I mean, imagine your plane drifting off center line to the right. If you erroneously try to correct by turning the wheel to the left, you will lower the right aileron into the windflow, causing the plane to swerve even farther to the right of center line. Result—one very befuddled pilot, with maybe a broken runway light and a bounding trip across the grass infield. (Should you ever wipe out a runway light, just bring the plane to a stop and count to ten. That's how long it takes the airport manager to race out in a little yellow pickup truck to collect the one-hundred twenty-five dollars that it costs to replace the darned thing.) The proper action to stop the swerve, naturally, is to correct with the rudder and nosewheel steering.

There is a time and place when you do intentionally *use* aileron drag. It is your first line of defense in combating a crosswind.

Correction for Crosswind

Let's conclude our discussion of directional control during the takeoff run with a few thoughts on crosswind correction. Nearly every takeoff requires you to make some correction for a crosswind—a totally windless day is rare; a wind straight down the runway is rarer still. Usually, there is a wind blowing against the vertical stabilizer that tries to "weathervane" the plane into the wind.

Your best defense against this crosswind on the takeoff run is deliberate, positive use of aileron drag. By turning the control yoke to deflect the *downwind* aileron downward, you create drag against the direction of the crosswind. Let's say that on this takeoff run, the crosswind blows from the left. Turn the control yoke in that direction and the right (downwind) aileron lowers to dig into the onrushing air. The crosswind tries to weathervane the plane to the left, but aileron drag counters with its right-hand tug to keep the plane astride the center line.

Now, a question arises as to how much aileron deflection should you use. Well, it is practically impossible to use too much aileron during the first moments of a takeoff run. The initial rolling speed is so slow that only minimal drag is created by the downward aileron. Therefore, begin the takeoff roll with full control deflection. Then, two things happen as your groundspeed increases. First, the aileron drag quadruples. Second, a crosswind affects a faster moving plane less than a slower moving one. (It is a combination of force and time.) As the rolling plane picks up speed, start decreasing the aileron deflection. Time this decrease so that near-zero aileron deflection exists just as you lift off, the next stage of your takeoff sequence.

■ Lifting Off

During the lift-off stage of the takeoff run, showcase two elements of control, precise airspeed control and accurate directional control, by effectively correcting for crosswind and compensating for P-factor.

Airspeed Control

Precise rotate speed is essential for best takeoff performance. If you try to lift off before the wings are ready to fly, you lengthen the takeoff run. Aerodynamic drag produced by the wings' too-early angle of attack is the villain here. If you try to hold the plane on the ground beyond the correct rotate speed, the takeoff distance lengthens also. Here, it is the rolling friction produced by the tires against the runway that slows the action.

Study the aircraft manual to determine the *exact* rotate speed recommended (Fig. 3-3). Apply back pressure on the yoke at this precise value to begin your lift-off.

Directional Control

Crosswind and the P-factor are two forces that try to disturb your directional control at lift-off. During the takeoff run, you corrected for the crosswind with deflected ailerons, which countered the weathervaning effect with aileron drag. At the moment of lift-off, however, this control deflection is reduced to near zero. Yet the crosswind still exists. You must maintain a crosswind correction if you are to climb out directly over the center line. So, establish a wind correction angle by making a shallow turn of a few degrees into the wind. (About 1 degree for each knot of crosswind component. For example, a 5-knot crosswind calls for about a 5-degree wind correction angle.)

P-factor, another element that enters the action to disturb your directional control during lift-off, results from the propeller's motion and the fact that the propeller blades are airfoils. As elevator pressure pitches the plane's nose upward, the right half of the propeller arc has a higher angle of attack than the left half (Fig. 3-4). With the right half of the propeller arc now producing excess thrust, the plane tends to swing to the left. Close attention to the runway's center line and quick rudder action, however, keeps the plane going straight.

PERFORMANCE

All airspeeds in this section are indicated airspeeds (IAS) except as noted and assume zero instrument error.

The performance data in this section has been established by flight tests and engineering calculation to assist you in operating your airplane. Flight tests were conducted under normal operating conditions using average piloting techniques with the airplane and engine in good condition. In using the following data, allowance for actual conditions must be made.

AIRSPEEDS (2150 POUNDS)

TAKE-OFF SPEEDS (0° Flaps)

Lift-Off	70 mph/61 kts
50 Ft	80 mph/70 kts

CLIMB SPEEDS

Cruising Climb Speed:
Full throttle	85 mph/74 kts

Best Rate-of-Climb Speed (5000 ft):
Flaps Up	83 mph/72 kts
Flaps Down (35°)	65 mph/56 kts

Best Angle-of-Climb (5000 ft):
Flaps Up	75 mph/65 kts

Fig. 3-3. A study of your aircraft's manual reveals the precise, recommended rotate speed.

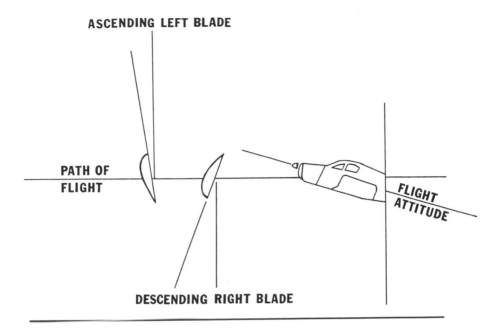

Fig. 3-4. When the nose pitches upward from the flight path, the right half of the propeller arc has a high angle of attack.

With airspeed increasing toward the best-rate-of-climb speed and directional control still firmly in hand, begin the next step in the takeoff sequence—the initial climbout.

■ Flying the Initial Climbout

The initial climbout stage of your takeoff sequence extends from lift-off through those moments that it takes to climb to pattern-departure altitude. There are four elements of precision flying that the examiner looks for during those moments: (1) airspeed control, (2) directional control, (3) traffic avoidance, and (4) correct pattern exit.

Airspeed Control

After lift-off, your plane will continue to accelerate toward *best-rate-of-climb speed*. This is the precise speed (stated in the aircraft operating manual) that the examiner wants to see—the speed that delivers the greatest gain in altitude for the *time* involved, consistent with engine cooling. Once this exact speed is reached, make any needed readjustment to the trim that holds this precise airspeed.

If the plane's manual directs you to use flaps for the takeoff, start retracting them one increment at a time. Retrim with each retraction so that airspeed is maintained at best rate throughout. This slow, orderly flap retraction provides a secondary benefit of preventing "dumping" the flaps. Pilots who dump all the flaps immediately after lift-off find that airspeed control is difficult to maintain. They may also suffer an actual loss of altitude while near the ground at a speed too close to the plane's flaps-up stalling speed.

Directional Control

The examiner wants to see precise directional control during the initial climbout. If allowed, crosswind or the left-turning force of torque can easily move you off course. At small rural airports, this can aim you toward a nearby hangar roof or line of trees. At busier airports, you might find that your first turn out of the pattern places you right in line of a faster plane that is departing behind you (Fig. 3-5). And traffic avoidance is a major concern around any airport.

Traffic Avoidance

As the old saying goes, a mid-air collision can spoil your entire afternoon. And there is no place where you are more apt to have a mid-air than around an airport (accident reports confirm the fact that most mid-airs occur at uncontrolled airports—those without control towers—during the takeoff or landing phases of the flight).

Let your traffic avoidance follow two lines of thinking: (1) spot traffic and (2) make yourself more visible to the other pilots. When you look for conflicting traffic, first search the strip of sky that lies 3 inches to either side of the horizon. Planes in that area are the planes *near* your altitude. Keep most of your attention on that particular band of sky, but do not neglect the rest. Search for conflicting traffic with a methodical, continuous cycle of observations—the sky ahead, the sky to the right and rear, the sky ahead, the sky to the left and rear. Between each cycle, make two more observations: (1) the flight and engine instruments (do the oil pressure or temperature gauges still register normal, for example, or does the heading indicator still show a course good and true?) and (2) the overall upper sky (is there a distant plane that *may* conflict in minutes to come?).

■ Leaving the Pattern

Traffic pattern exits are standardized. This is necessary so that departing and arriving pilots will know what to expect of one another. The

Fig. 3-5. Directional control during the initial climbout is important.

two acceptable methods for departing the pattern are (1) to continue straight out on the runway heading and (2) to execute a 45-degree turn after reaching the traffic pattern altitude.

If executing a 45-degree departure, make that turn to either the left or the right in accordance with the runway's established traffic flow; left-hand traffic flows are standard. However, many runways employ a non-standard right-hand flow. Any non-standard flow is noted in the FAA *Airport/Facility Directory.* Prudent pilots own a copy of this booklet, which describes services available and procedures to follow at each airport in the country.

At controlled airports the tower is usually quite willing (traffic conditions permitting) to sanction a non-standard course that would more quickly put you on your desired course. Remember, however, that you must get their permission before straying from the standard procedures.

With your pattern exit established, you are ready to start the sixth step of the takeoff sequence—climbing to cruise altitude.

■ Climbing to Cruise Altitude

Your flight-test examiner will look for five specific elements in your climb to your cruising altitude:

1. Traffic awareness.
2. Precise airspeed control.
3. Precise heading control.
4. Proper powerplant operation.
5. Accurate level-off at cruise altitude.

Traffic Awareness

Let alertness for traffic continue to be foremost in your mind as you climb toward cruising altitude. You are in an area for high traffic probability as you depart an airport. Think of that airport as the hub of a wheel with spokes as paths of arrivals leading to it. Then, think of your traffic avoidance efforts in terms of spotting traffic in a timely manner and helping *unseen* traffic to spot *you*.

If you are a bit on the short side in height, or are flying behind a large engine, you may find forward visibility restricted by a cowl that can easily hide straight-ahead traffic when in the climb attitude. In this case it would be wise to clear the forward area for head-on traffic every several seconds during the climb. Either make a shallow S-turn of 20 degrees to either side of the flight path or momentarily lower the nose for a quick look-see ahead. Make the most efficient use of your eyes.

Remember, your eyes work like a telescope. You need to vary your range of focus. There is a tendency to stay focused on the horizon, but that focus can look right through a nearby plane. Also, as you search the overall sky for traffic, pay particular attention to that band of sky 3 inches above and below the horizon for the planes that are sharing your altitude. Refrain from sweeping your head from side to side as you look for these planes. Rather, turn your head to each sector of sky in turn; then, move your *eyes* instead of your head for an unblurred look.

Monitor the appropriate local radio frequency (tower; approach/ departure; common traffic advisory frequency) until you are several miles away from the airport. You will hear the position reports and be better able to pick up the traffic. However, no matter how hard you look, you will not spot *all* the traffic — believe me. Therefore, as you search for planes, take actions that alert *your* presence to unseen traffic. (And as long as *one* pilot sees the other plane, there will be no collision unless that other pilot harbors some weird death wish.) Here are five basic actions that you might want to put into play to attract attention:

1. Even during daylight departures, keep your landing light ablaze until you are well clear of the traffic area. Head-on traffic is the toughest to spot, but with a landing light aimed at them, *they* are sure to spot *you.*

2. Wave at them by giving your wings a waggle a couple of times a minute during the climb. This action draws their attention, like waving a red flag at a bull. (No need to explain this move to your flight-test examiner. That examiner knows what you're doing and silently thanks you for the consideration.)

3. Even if you are not working with ATC (Air Traffic Control), keep your transponder winking its heart out. A controller will see your blip on the radar and will warn other pilots of your presence. (And trust me — there are many times when *neither* pilot sees the other. In these instances a good controller can provide the necessary third set of eyes.)

4. If your plane has strobes, keep them flashing. They are highly visible, even in bright sunlight. (A red rotating beacon, on the other hand, is not too useful during daylight. You usually spot the plane before you see the beacon.)

5. When departing a non-tower airport, give an altitude and position report at each 1000 feet of height until you reach cruising altitude. Broadcast these reports on the airport's designated common traffic advisory frequency (CTAF).

Precise Airspeed Control

The flight-test examiner expects to see close adherence to the plane's recommended *best-rate-of-climb speed* during the climb to cruising altitude. Best-rate-of-climb speed is not an arbitrary figure. It is the precise airspeed that delivers the greatest gain in altitude for a given *time* period and that is consistent with adequate engine cooling. The manufacturers have done considerable testing to determine the best-rate-of-climb speed for their planes, and state the value in no uncertain terms within the operational manuals for each plane (Fig. 3-6). Hit the speed right on the mark. If you climb at a speed either below or above this mark, you

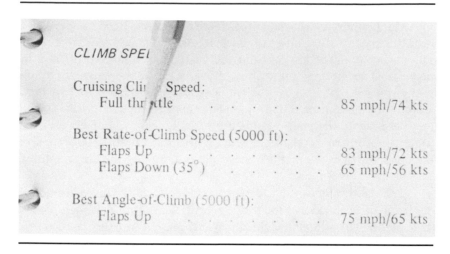

CLIMB SPEI

Cruising Clir Speed:
  Full thr tle   . .   .   .   .   .   85 mph/74 kts

Best Rate-of-Climb Speed (5000 ft):
  Flaps Up   .   .   .   .   .   .   83 mph/72 kts
  Flaps Down (35°)   .   .   .   .   65 mph/56 kts

Best Angle-of-Climb (5000 ft):
  Flaps Up   .   .   .   .   .   75 mph/65 kts

Fig. 3-6. Aircraft manuals state the exact best-rate-of-climb speed to use.

prolong the time to reach altitude. You will also contribute to engine overheating—excess time at full bore leads to excess engine heat, which leads to excess engine wear. The examiner does not want to see this happen.

Hitting the best-rate-of-climb speed and holding it depends on three simple things:

1. Knowing the *correct,* specific climb speed. This one's easy—the precise speed is stated in the manual. Using this value rather than the pass-along advice of another pilot is the only way to go. The manufacturer knows what works best for that airplane.

2. Knowing the *precise* pitch attitude that delivers the best-rate airspeed. Knowing this attitude means observation, observation, observation! In each of your training climbs to altitude, observe how the horizon lies relative to the top of the engine cowl. (If you *really* want to fix this attitude in your mind, here's a tip to help you. While maintaining the best rate of climb, continue to sit back in your normal seat position. While doing so, reach forward with the eraser end of a pencil and smudge the windscreen right where the horizon line crosses it. You will then find how easy it is to return to that attitude on demand.)

3. Knowing how to *trim* the pitch to the needed attitude. Knowing how to use the trim control takes practice—and a lot of it. Many fledgling pilots fail to subject themselves to this constant trim-control practice. They are content to fly the plane out of trim and then just muscle

the flight controls to override their out-of-trim airplane. As a result, they never become smooth pilots.

Don't fall into that trap. Practice trim control on each flight whenever it must be readjusted. (Trim must be readjusted anytime the pilot changes either airspeed, flap settings, or power—even slightly. Both factors disturb the plane's pitching moment, making trim adjustment necessary.) By trimming each and every time the plane needs it, you soon become a pro with the trim wheel—holding the best-rate-of-climb speed becomes a snap.

Precise Heading Control

One primary aerodynamic factor that tries to turn you away from a precise climbout heading is torque. From the discussion in chapter 1, you know that torque will try to turn your plane to the left throughout the climb. You can compensate for this in one of two ways, depending on how your plane is equipped. If your plane has rudder trim control, then adjust the trim control for straight-ahead flight. If your plane does not have this feature, then just hold enough right-rudder pressure to keep the nose from turning.

Proper Powerplant Operation

Throughout the flight test, your examiner looks for good engine-operation procedures. No other maneuver showcases these procedures better than the climb to cruise altitude. Fit your powerplant-management efforts into three categories: (1) monitoring engine gauges for proper powerplant performance, (2) maintaining the manufacturer's recommended power setting, and (3) managing the fuel flow.

ENGINE GAUGES. During the climb, let your eyes scan the engine gauges (ammeter, oil pressure, engine temperature, and tachometer) as you pass through each 1000 feet on the altimeter. If you see an improper indication of any of these instruments, bring the discrepancy to the examiner's immediate attention. (*Don't* wait until the examiner brings it to your attention with a pencil tap to the offending instrument.) If the malfunction appears non-critical, the examiner may ask you to perform the standard discrepancy checks of the instruments at fault.

Ammeter. If the ammeter shows a discharge or its warning light goes on, the alternator switch may need to be reset. To do so, simply turn off the *alternator half* of the master switch for a moment. If the warning light continues to burn after returning the switch to the "on" position, recycle the process. If recycling doesn't fix the problem, your alternator is probably shorting. Then, turn off the master switch to conserve battery power until you need the radio for landing. (The engine will not quit. It is fired by magneto power, independent of the electrical system.)

Oil Pressure. If the oil pressure is falling, your time aloft may be limited. Tell your examiner the flight test is over for the day, and head *directly* back to the airport.

Engine Temperature. It is not uncommon for an engine to heat up a bit during a prolonged climb. Usually, however, the instrument needle stays within the upper reaches of the green arc. Should engine heat become a concern during your climb, take one of two actions depending on how your plane is equipped. If your plane is equipped with cowl flaps, open them for better cooling. (Actually, most aircraft manuals call for cowl flaps before the climb begins.) If your plane is not equipped with cowl flaps, level from the climb and reduce throttle to cruise power. If the overheating was due to an unusually hot day, a few minutes in cruise flight should cool the engine. If, however, your efforts at cooling the engine are to no avail, the high temperature may be a symptom of powerplant trouble brewing. In this case, you would be wise to head for your airport and let a mechanic render an opinion.

Tachometer. Most flight tests for the private pilot certificate are flown in light trainers equipped with fixed props. Usually, the flight manuals for these planes specify that you are to climb to cruise altitude at full power.

POWER SETTING. Maintaining full power throughout a lengthy climb may require some action on your part. Throttle creep, carburetor ice, and improper fuel mixture are three negative factors that may diminish your engine's output during the climb.

Throttle Creep. In light aircraft, throttles have a habit of not staying in position, even when the throttle lock is used. They creep. (I sometimes feel that manufacturers employ a vice-president in charge of throttle creep, just to make sure it happens.) So, at each 1000 feet of climb, check your tachometer for a full-power indication. If you see a power loss, let your first discrepancy check confirm that the throttle has not changed position while your throttle hand was removed from the control. (Prudent pilots keep a hand on the throttle as much as possible.)

Carburetor Ice. It is wrong to think of carburetor icing as something that happens only at reduced-power settings, when the engine cylinders are producing minimal heating around the carburetor. Carburetor ice can form even at full climb power, if the air's moisture content is high enough.

To better understand how carburetor ice can form, look at the inside of a carburetor (Fig. 3-7). The primary job of the carburetor is to mix air and gasoline to form a fuel that the cylinders can utilize. Follow the airflow to see how ice can occur in the carburetor and the extent to which it reduces the required amount of fuel, thus producing restricted engine power.

First, the outside air (containing moisture) enters the carburetor. It then passes through a narrow throat to gain velocity with which to reduce atmospheric pressure (faster-flowing air produces lower pressure). This lowered pressure lets the air suck gasoline from the carburetor's fuel jet. Finally, the air is decompressed in a larger expansion chamber. This sudden expansion cools the air. (That's how a refrigerator works.) In a carburetor, that expanded air can cool 60° below its original, outside-air temperature. Even on a hot, humid day the temperature inside the carburetor can chill down below freezing. The moisture carried in by the air

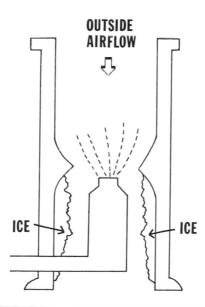

Fig. 3-7. Rapid cooling through sudden expansion can create ice in the lower chamber of the carburetor.

then freezes to form a layer of ice on the interior surfaces of the carburetor. If allowed to continue, this build-up of ice can seriously restrict the fuel flow necessary for climb power.

If you note a drop in climb power and you have already ruled out throttle creep, run a discrepancy check for carburetor ice. Apply carburetor heat. When you apply carburetor heat, you duct a blast of heat from the exhaust stacks to preheat the air going to the carburetor.

If there is ice in the carburetor, you will note three indications on your tachometer (or on your manifold pressure gauge if you are flying a plane equipped with a constant-speed propeller):

1. A power reduction when heat is first applied (caused by the hot air).
2. An increase in power as the ice is eliminated (the engine may hiccup as it tries to digest some water).
3. Another increase in power as you turn off the heat (as the flow of hot air ceases).

Improper Fuel Mixture. Light-aircraft engines are designed to operate with a fuel mixture of 1 part gasoline to 15 parts air (1 pound of gasoline to 15 pounds of air). When the amount of air passing through the carburetor exceeds this 1:15 ratio, the mixture is too lean. When too lean, the fuel burns too rapidly in the cylinders and produces *detonation*, which results in engine roughness and reduced power.

Too much gasoline for the air available, on the other hand, results in an over-rich mixture, which results in unburned fuel within the cylinders. This unburned fuel collects against the cylinder walls to form "hot spots" of carbon, along with spark plug fouling. These hot spots actually ignite the incoming fuel before the spark plugs can and before the piston reaches the top of its power stroke. This *preignition* results in a rough-running engine and a loss of power.

An over-rich mixture for a short duration does, however, hold one advantage for the engine. It helps cool the engine during times of peak power. This is why you use a full-rich mixture for takeoff and for the first several thousand feet of climb; you are using full throttle with the hope of producing all the power the engine can produce. A rule of thumb says, lean the mixture when the engine is producing no more than 75% power.

If your climb to cruise altitude extends above 5000 or 6000 feet above sea level (MSL), you may find it advantageous to lean the mixture even during the full-throttle climb. Two things occur at or about this higher altitude. First, the air is thinner. Without leaning the flow of gasoline to match this thinner air, a full-rich mixture exceeds the 1:15 desired fuel ratio. The engine may run rough and not deliver its best power. Second, at these higher altitudes the engine may produce no more than 75% power, even at full throttle; therefore, leaning is appropriate. (The typical small engine with a fixed prop loses about 4% of its power with each 1000 feet of altitude.)

If your engine starts to run rough high into your climb, leaning could be in order. This brings us to the next question, How do we manage the fuel flow?

MANAGING THE FUEL FLOW. When managing the fuel flow (gasoline and air mixture) to attain the 1:15 ratio the pilot is concerned with two elements — the amount (by weight) of gasoline going to the carburetor and the amount (by weight) of the air allowed to pass through the carburetor. The pilot has two controls, the mixture control and the throttle, available with which to manage these two elements (Fig. 3-8).

Fig. 3-8. The pilot manages the fuel flow with the mixture control and the throttle.

The mixture control regulates the amount of gasoline entering the carburetor via the fuel jet (Fig. 3-9). Move the mixture control forward and more gasoline enters the jet; retard (pull back) the control and less gasoline flows.

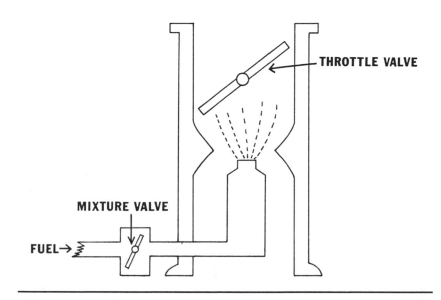

THROTTLE VALVE

MIXTURE VALVE

FUEL→

Fig. 3-9. The mixture control regulates the amount of fuel entering the carburetor. The throttle manages the volume of air.

The throttle regulates the amount of air allowed to pass through the carburetor, by means of a simple butterfly valve (Fig. 3-9). Advance the throttle and the valve opens wider; retard the throttle and the valve opening reduces. The pilot uses a combination of these two controls to attain and maintain a usable fuel mixture ratio.

In a climb, most light trainers run at full throttle to get all the power the engine can deliver. During takeoff and the climb through the lower altitudes, a full-rich mixture is appropriate; the engine develops in excess of 75% power. But if your climb extends above 5000 or 6000 MSL (above sea level), fuel-leaning might be in order. At about this altitude, the air is so thin that a full-rich mixture delivers a fuel that exceeds the 1:15 ratio to the point where the engine runs rough. At this point of roughness due to thin air, you can be sure that the engine is producing less than 75% power.

Leaning the mixture is simplicity itself and is the same procedure whether in a climb or level cruise flight. To lean the fuel mixture, slowly retard the mixture control. The first things you will notice are that the engine runs smoother and the tachometer (on a fixed-pitch-prop-equipped engine) gains revolutions per minute (rpm) as you get closer to the optimum 1:15 fuel ratio. Keep retarding the mixture control slowly until the engine hiccups or the tachometer shows a small loss of power. Stop leaning the mixture at this point of engine roughness. When you "lean to roughness," *one* cylinder is too lean and starts to misfire. At this point, slightly advance the mixture control to smooth out the engine and pick up a few rpm on the tachometer. Now you have the leanest cylinder at a perfect 1:15 ratio, with the remaining cylinders *very* slightly rich — and you have a fine-running engine.

Let me say this — some fledgling pilots are reluctant to retard the mixture to the point of engine roughness, because they are worried that they might cause the engine to quit. But, with slow movement of the mixture control, this just won't happen. The engine roughness is normal, and it won't stop running. Of course, if you *yank* the mixture control *back to its stop,* the engine *will* quit from fuel starvation. But even then, if you erroneously do so, a forward push on the control will have the engine (with its still-windmilling prop) running again. No problem. If you need convincing on this matter, let me relate a common training error to you.

About half the students I train experience the mistake. Sooner or later on a landing approach, they mistakenly pull the mixture control to its stop, rather than the carburetor heat control, which is what they really meant to grab. At this moment, the engine stops and they usually look over at me with an expression that asks, Huh? Then I point to their mistake, they see it and correct the matter, and the windmilling prop refires the engine right back to where the action started. Invariably, they finally respond with an embarrassed giggle and a heartfelt promise not to do it again and that ends the matter. No problem.

I remember a student who made this mistake, except we were flying at minimum controllable airspeed. Here, the airspeed wasn't sufficient to keep the prop windmilling. The engine stopped and *stayed* stopped, even after the mix-up of carburetor heat and mixture control was corrected. We were suddenly aboard a glider. The student's fidgeting seemed to be asking for advice — so I advised: Take a moment to think it out and let me know what you decide. (You *can* get a stopped prop windmilling again with a steep, fast dive. But I hoped that solution would not occur to the student — I like my flying on the sedate side.) A few seconds later, the student exclaimed, Aha, the starter! So a twist of the key refired the engine and we flew on our merry way. Again, no problem. The point of this episode (if there is a point) is simply, Don't view a simple mistake in the cockpit as the opening jaws of doom. Rather, when you discover a mistake on your part, just take a moment to ask yourself, Now what the heck should I logically do? The answer is usually simple common sense. Flying just isn't all that complicated.

Once you have demonstrated good climb procedures, you have only one remaining segment of the takeoff sequence to demonstrate—a perfect level-off to cruise altitude.

■ Leveling to Cruise Altitude and Airspeed

Complete your takeoff and climb with a precision level-off to a precise altitude and airspeed—plan to hit cruise speed just as you level the plane at your exact selected altitude. It's easy to do with four simple steps.

1. Allow yourself an unrushed level-off. Start lowering the plane's nose *slowly* toward the level attitude, 50 feet below your desired altitude for each 500 feet per minute (or 10 feet for each 100 feet) on the vertical speed indicator. Thus, a 400-feet-per-minute climb asks for a 40-foot lead; 600 feet per minute calls for a 60-foot lead. Give quick glances at the upward-climbing altimeter, to time a pitch reduction that gains the altitude just as the plane levels.

2. Accelerate quickly to cruise speed by leaving the throttle at climb power until cruise speed is attained. Once you reach cruise speed, *then* promptly reduce throttle to exact cruise power. If you reduce power before this instant, you will lose some altitude. If you delay the power reduction, you will climb above your desired altitude.

3. Keep the controls coordinated throughout the leveling process—hold right rudder against torque during the climb, but keep the ball caged during level-off, by slowing releasing that rudder pressure as the plane accelerates to cruise speed and the throttle reduces to normal cruise power.

4. Make the final transition to cruise flight, by fine-tuning the power setting, leaning the mixture to its correct cruise fuel flow, and trimming the controls to fingertip pressure.

If you let your demonstration of takeoff and climb play out with the seven stages of the maneuver—each stage with its own achievements and goals—you will most likely elicit a satisfied grunt from the examiner. But don't expect a rewarding smile. After all, the examiner *expects* to see a job well done.

SHORT-FIELD TAKEOFF AND DEPARTURE

If adventure truly calls you to flying, then short-field operations are probably in your future. The most interesting places in our land, it seems to me, are served by the shortest runways on the sectional chart.

Your flight examiner wants to know that your skills are up to the task. Basically, your examiner is looking for an understanding of the airplane flight manual's takeoff performance chart, which allows an evaluation of the takeoff environment, and piloting skills that launch the plane with a minimum takeoff run and the most efficient initial climbout.

■ Aircraft Flight Manual

Every student pilot should possess a flight manual for the plane they fly. And, the manual must be studied carefully.

Thorough study is needed for pilot safety. There is no such thing as a forgiving airplane. Every plane has operating limits, and every plane must be flown in a prescribed manner if the pilot is to remain free of accidents. In light planes, especially, the parameters of these limits are quite narrow in terms of loading, speeds, power, and aerodynamic forces. Operate the plane within these limits and the plane will perform just as the manufacturer predicts. Operate *beyond* these limits, however, and you are suddenly a new test pilot engaged in on-the-job training. Manufacturers state their tested and proven procedures with the express purpose of keeping pilots within the plane's designed limits of operation.

If you have not *thoroughly studied* your aircraft manual before presenting yourself for the certification flight, I urge you to do so. You will discover a free hour of the best instruction you have ever received from the best flight instructor. Knowing and believing your flight manual enhances your level of safety — one that will follow you through your flight career. Throughout your flying lifetime, you will be plagued by the admonishments of well-meaning fellow pilots, trying to tell you how your plane should best be flown. I'm convinced many have never read the manual. Suggestions often range from the outrageous (and hazardous) to the comic (opening the door to embarrassment). Know and follow your manufacturer's tested and recommended procedures. It is the *only* way to go. Don't succumb to the pass-along advice of others. Look to three sections of the aircraft manual for the manufacturer's advice about short-field takeoffs — the operations, aircraft loading, and the takeoff and climb-performance charts.

Operations

The operations section often spells out special procedures to use when confronting a short-field takeoff. Quite often, normal takeoff procedures such as flap extension, initial climb speed, or special precautions are modified.

Aircraft Loading

The takeoff performance that the manufacturers' takeoff and climb charts predict assume that recommended pilot procedure and stipulated loading, within allowed gross weight and within balance, are strictly adhered to. A plane that is launched overweight or out of balance has such unpredictable flying characteristics that even the manufacturer has no idea what will happen.

The loading charts and graphs of most light aircraft are easy to interpret. Others are not. Whichever is the case for the plane you fly, I suggest that you have your instructor guide you through several sample loading problems before you present yourself for the flight test. And by all means, base one of your examples on the actual weights of both you and your instructor. If you fly the smallest trainers and each of you equals my own weight, you may be in for a surprise. Let me explain.

The trainer I fly is a small, "garden-variety" two seater—a plane that is said to be forgiving. Yet, with a student aboard who weights the same as I do and with "topped" tanks, the plane is unsafe to fly. It is slightly over the maximum allowable gross weight, and with both seats shoved back to accommodate big feet, it is slightly out of balance aft. If the plane were to stall in this configuration, recovery would be in question. (How do I handle these students? I train them in a larger airplane.)

Takeoff and Climb-Performance Charts

The takeoff and climb tables calculate performance against seven elements that affect the distance needed for a safe departure: (1) obstacle clearance distance, (2) field elevation, (3) outside air temperature, (4) aircraft loaded weight, (5) flaps and airspeed, (6) headwind component, and (7) runway surface. Look at the performance tables for your plane to see how each is charted.

A comment or two about each of these elements may help you respond to some of the examiner's questions concerning departure performance.

OBSTACLE CLEARANCE DISTANCE. The takeoff performance chart (Fig. 3-10) provides distance figures for the ground run and for the total

4-6

Sport B19

TAKE-OFF DISTANCE — HARD SURFACE

ASSOCIATED CONDITIONS

POWER	FULL THROTTLE
MIXTURE	LEAN TO MAXIMUM RPM, THEN ENRICH SLIGHTLY
FLAPS	UP
RUNWAY	LEVEL, DRY, HARD SURFACE
WEIGHT	2150 LBS

TAKE OFF SPEEDS

LIFT OFF	70 MPH/61 KTS
50 FT	80 MPH/70 KTS

WIND COMPONENT DOWN RUNWAY KNOTS	SEA LEVEL			2000 FT			4000 FT			6000 FT			8000 FT		
	OAT °F °C	GROUND ROLL FEET	TOTAL OVER 50 FT OBSTACLE FEET	OAT °F °C	GROUND ROLL FEET	TOTAL OVER 50 FT OBSTACLE FEET	OAT °F °C	GROUND ROLL FEET	TOTAL OVER 50 FT OBSTACLE FEET	OAT °F °C	GROUND ROLL FEET	TOTAL OVER 50 FT OBSTACLE FEET	OAT °F °C	GROUND ROLL FEET	TOTAL OVER 50 FT OBSTACLE FEET
0	23 -5	836	1331	16 -9	953	1510	9 -13	1089	1717	2 -17	1247	1956	-6 -21	1430	2237
	41 5	930	1478	34 1	1062	1680	27 -3	1215	1913	20 -7	1393	2182	13 -11	1600	2495
	59 15	1034	1635	52 11	1178	1861	45 7	1350	2123	38 3	1550	2425	31 -1	1784	2777
	77 25		1803	70 21	1303	2055	63 17	1496	2347	56 13	1719	2686	49 9	1981	3079
	95 35		1982	88 31	1435	2262	81 27	1649	2586	74 23	1899	2963	67 19	2191	3402
15	23 -5	642	1195	16 -9	738	1361	9 -13	850	1652	2 -17	980	1773	-6 -21	1132	2030
	41 5	717	1331	34 1	826	1517	27 -3	953	1733	20 -7	1100	1984	13 -11	1274	2274
	59 15	799	1476	52 11	921	1685	45 7	1063	1928	38 3	1230	2218	31 -1	1426	2538
	77 25	886	1631	70 21	1023	1865	63 17	1182	2136	56 13	1370	2462	49 9	1590	2826
	95 35	979	1796	88 31	1131	2056	81 27	1310	2359	74 23	1520	2711	67 19	1766	3122
	23 -5	471	1084	16 -9	548	1232	9 -13	638	1415	2 -17	743	1621	-6 -21	867	1860
	41 5	530	1209	34 1	617	1383	27 -3	720	1584	20 -7	840	1817	13 -11	982	2089
	59 15	594	1344	52 11	693	1539	45 7	808	1765	38 3	945	2028	31 -1	1105	2335
	77 25	663	1488	70 21	774	1706	63 17	904	1959	56 13	1057	2256	49 9	1239	2600
	95 35	738	1642	88 31	861	1885	81 27	1007	2167	74 23	1180	2498	67 19	1384	2884

Fig. 3-10. The longer "total distance to clear obstacle" is of more interest to the departing pilot.

distance needed to clear a 50-foot obstacle. It is this second, longer distance that is of interest to the pilot, even if no obstacles are present, for you want to be *at least* that high before you cross the airport boundary.

If obstacles *are* present, you may need to modify the charted obstacle-clearance distance upward, because, alas, real obstacles seldom come in FAA standard 50-foot heights. Modify the chart's required distance for higher obstacles with three steps:

1. Subtract the *ground run* distance from the *total distance* to obtain the *climb distance* needed for each 50 feet of obstacle height. (The chart lying open before me as I write this chapter, for example, charts a ground run of 630 feet with a total-to-clear-obstacle distance of 1095 feet. By subtracting the little number from the big number, I arrive at 465 feet of climb distance required for each 50 feet of obstacle height.)

2. Estimate the obstacle's height. (If, for instance, the obstacle confronting me was a Florida long-leaf pine, I'd estimate its height at about 100 feet.)

3. Add the climb distance for each 50 feet of the estimated obstacle height in excess of the chart's standard 50-foot obstacle. (For my pine tree, I'd add another 465 feet for a calculated obstacle clearance distance of 1550 feet.)

FIELD ELEVATION. Field elevation plays a major role in determining the required takeoff distance (Fig. 3-11). Air density is the culprit here.

TAKE-OFF DISTANCE — HARD SURFACE

ASSOCIATED CONDITIONS

POWER   FULL THROTTLE
MIXTURE   LEAN TO MAXIMUM RPM, THEN ENRICH SLIGHTLY
FLAPS   UP
RUNWAY   LEVEL, DRY, HARD SURFACE
WEIGHT   2150 LBS

TAKE OFF SPEEDS

LIFT OFF   70 MPH/61 KTS
50 FT   80 MPH/70 KTS

WIND COMPONENT DOWN RUNWAY KNOTS	SEA LEVEL OAT °F / °C	GROUND ROLL FEET	TOTAL OVER 50 FT	2000 FT AT / °C	GROUND ROLL FEET	TOTAL OBSTACLE FEET	4000 FT OAT °F / °C	GROUND ROLL FEET	TOTAL OBSTACLE FEET	6000 FT OAT °F / °C	GROUND ROLL FEET	TOTAL OBSTACLE FEET	8000 FT OAT °F / °C	GROUND ROLL FEET	TOTAL OVER 50 FT FEET
0	23			16 / 0	953	1510	9 / 13	1089	1717	2 / 17	1247	1966	6 / 21	1430	2232
	41	1062		34 / 1	1062	1680	27 / 3	1215	1913	20 / 7	1393	2182	13 / 11	1600	2495
	59	1153		52 / 11	1178	1861	45 / 7	1350	2123	38 / 13	1550	2425	31 / 1	1784	2777
	77			70 / 21	1303	2055	63 / 17	1495	2347	56 / 13	1719	2686	49 / 9	1981	3079
	95	1302		88 / 31	1435	2262	81 / 27	1649	2686	74 / 23	1899	2963	67 / 19	2191	3402
15	23			16 / 0	738	1361	9 / 13	860	1552	2 / 17	980	1773	6 / 21	1132	2030
	41	1331		34 / 1	826	1517	27 / 3	953	1733	20 / 7	1100	1994	13 / 11	1274	2274
	59	1476		52 / 11	921	1685	45 / 7	1063	1928	38 / 13	1230	2210	31 / 1	1426	2538
	77	896	1631	70 / 21	1023	1865	63 / 17	1182	2136	56 / 13	1370	2452	49 / 9	1590	2820
	95	979	1796	88 / 31	1131	2056	81 / 27	1310	2359	74 / 23	1520	2711	67 / 19	1766	3122
	5	471	1084	16 / 0	548	1232	9 / 13	638	1415	2 / 17	743	1621	6 / 21	867	1850
	5	530	1209	34 / 1	617	1383	27 / 3	720	1584	20 / 7	840	1817	13 / 11	982	2089
	15	594	1344	52 / 11	693	1539	45 / 7	808	1765	38 / 3	945	2028	31 / 1	1105	2335
	25	663	1488	70 / 21	774	1706	63 / 17	904	1959	56 / 13	1057	2256	49 / 9	1239	2600
	95	738	1642	88 / 31	861	1885	81 / 27	1007	2167	74 / 23	1180	2498	67 / 19	1384	2884

Fig. 3-11. Higher field elevations require greater takeoff distances. Notice how the distances increase for 2000- to 4000-foot elevations.

At higher elevations the reduced atmospheric pressure allows the molecules of air to spread out, resulting in "thin" air. So, in order to get the required number of molecules of air flowing over the wing, the plane must accelerate to a faster true rotate speed. Accelerating to this faster true airspeed requires a longer ground run.

Once airborne, the plane also needs a greater climb distance to clear the obstacle. The rate of climb suffers in thinner air, primarily because the engine's output is reduced as it takes in this thin air.

There is another reason, often overlooked, that takeoff and climb distance is lengthened at high field elevations — the difference between *indicated airspeed* (IAS) and *true airspeed* (TAS). The thinner air of high altitude affects the air's impact on the airspeed indicator's pitot tube in the same manner that it affects the airflow across the wing. You must fly faster to get the same impact. This means that at higher elevations your plane is actually flying faster than shown on the airspeed indicator. To estimate true airspeed, add 2% to the indicated airspeed for each 1000 feet of elevation. (At 5000 feet, for example, a plane indicating 100 knots is actually flying about 110 knots true airspeed.)

But, remember, that the aircraft's manual will recommend *sea-level-* indicated rotate and climbout speeds, even when executing a high-elevation takeoff. This recommendation is valid because the thin air affects both the airspeed indicator and the wings in the same manner. So, if the manual recommends 60 knots for lift-off at sea level, use that same rotate speed for your mountain-airport departure. Just be aware that the 60 knots IAS actually has you moving at about 66 knots TAS, and you'll need a longer ground run before lift-off. Similarly, if the manual states a sea-level climbout speed for obstacle clearance of 65 knots IAS, use that same indication at high elevations. The TAS will then be about 71 knots. And traveling at this faster speed across the ground requires additional distance with which to gain the altitude needed to clear the obstacle.

OUTSIDE AIR TEMPERATURE. High temperature has the same detrimental effect on takeoff performance as does high elevations — and for the same reason. Heat drives the molecules of air farther apart, thus, the wings and engine have to work harder.

Many takeoff charts calculate performance only against *standard* temperature for the field elevation in question (Fig. 3-12). In this case you must convert the figures given, to allow for the *actual* temperature. The rule of thumb for this conversion is, increase the stated takeoff distance by 10% for each 20°F above "standard" for the field elevation. (Refer to the outside air temperature gauge mounted in the windscreen to determine the existing temperature.)

AIRCRAFT LOADED WEIGHT. Aircraft loading seriously affects take-off performance. The more your airplane weighs, the greater the inertia the engine must overcome to accelerate the plane to rotate speed.

TAKE-OFF DISTANCE — HARD SURFACE

ASSOCIATED CONDITIONS:

POWER	FULL THROTTLE
MIXTURE	LEAN TO MAXIMUM RPM, THEN ENRICH SLIGHTLY
FLAPS	UP
RUNWAY	LEVEL, DRY, HARD SURFACE
WEIGHT	2150 LBS

TAKE OFF SPEEDS:
LIFT OFF: 70 MPH/61 KTS
50 FT: 80 MPH/70 KTS

WIND COMPONENT DOWN RUNWAY KNOTS	SEA LEVEL				2000 FT				4000 FT				6000 FT				8000 FT			
	OAT F	C	GROUND ROLL FEET	TOTAL OVER 50 FT OBSTACLE FEET	OAT F	C	GROUND ROLL FEET	TOTAL OVER 50 FT OBSTACLE FEET	OAT F	C	GROUND ROLL FEET	TOTAL OVER 50 FT OBSTACLE FEET	OAT F	C	GROUND ROLL FEET	TOTAL OVER 50 FT OBSTACLE FEET	OAT F	C	GROUND ROLL FEET	TOTAL OVER 50 FT OBSTACLE FEET
0	23	-5	836	1331	16		953	1510	9	-13	1089	1717	2	-17	1247	1966	-6	-21	1430	2232
	41	5	930	1478	34	1	1062	1690	27	-3	1216	1913	20	-7	1393	2182	13	-11	1600	2496
	59	15	1030	1636	52	11	1178	1861	45	7	1350	2123	38	3	1550	2425	31	-1	1784	2777
	77	25	1137	1803	70	21	1303	2055	63	17	1495	2342	56	13	1719	2686	49	9	1981	3079
	95	35	1251	1982	88	31	1435	2262	81	27	1649	2586	74	23	1899	2963	67	19	2191	3402
15	23	-5	642	1195	16	9	738	1361	9	13	850	1552	2	-17	980	1723	-6	-21	1132	2030
	41	5	717	1331	34		826	1517	27	3	953	1733	20	-7	1100	1984	13	-11	1274	2274
	59	15	799	1476	52		921	1685	45	7	1063	1928	38	3	1230	2210	31	-1	1426	2538
	77	25	886	1631	70		1023	1865	63	17	1182	2136	56	13	1370	2452	49	9	1590	2820
	95	35	979	1796	88		1131	2056	81	27	1310	2359	74	23	1520	2711	67	19	1766	3122
30	23	-5	471	1084	16		48	1237	9	13	638	1415	2	-17	743	1621	-6	-21	867	1860
	41	5	530	1209	34		7	1383	27	3	720	1584	20	-7	840	1817	13	-11	982	2089
	59	15	594	1344	52			1539	45	7	808	1765	38	3	945	2028	31	-1	1105	2335
	77	25	663	1488	70	2	1706		63	17	904	1959	56	13	1057	2255	49	9	1239	2600
	95	35	738	1642	88	31	1885		81	27	1007	2167	74	23	1180	2498	67	19	1384	2884

Fig. 3-12. In this example, the performance chart offers several representative temperatures. If your handbook only contemplates "standard temperature" you should modify the distances required by the text's rule of thumb.

Ground run is lengthened. Once airborne, rate of climb suffers as the engine pulls its load "uphill."

Most takeoff charts state required distances at various gross weights. However, some of the charts for the lightest planes do not. These charts often state the required distance needed at the plane's *maximum allowable* gross weight. If this chart presentation applies to the plane you fly, the rule of thumb is, for lightly loaded takeoffs, reduce the charted distance by 5% for each 100 pounds below maximum allowable gross weight.

FLAPS AND AIRSPEED. Manufacturers clearly state the flap position that best launches their product (Fig. 3-13). Often the manual states two flap settings: one for normal departures, the other for short-field takeoffs over an obstacle. Use the setting appropriate to the situation.

Nearly all manuals recommend a *slow* flap retraction (by increments) *after* a safe altitude and airspeed is attained. This is important in that rapidly "dumping" the flaps ordinarily results in an erratic climb speed and an actual loss in altitude. Additionally, the speed recommended for obstacle clearance is usually close to the flaps-up stall speed.

Most aircraft manufacturers recommend a normal rotate speed when departing from a short runway, along with normal *best-rate*-of-climb speed to cruise altitude once the obstacle is cleared. But, nearly all aircraft manuals call for a particular climb speed from rotation to obstacle clearance. This airspeed is defined as *best-angle*-of-climb speed; the speed that gains the greatest *altitude* in the shortest *distance*. This airspeed is included in the takeoff chart of nearly all models, often labeled

4-6

Sport B19

TAKE-OFF DISTANCE — HARD SURFACE

ASSOCIATED CONDITIONS:

POWER	FULL THROTTLE
MIXTURE	LEAN TO MAXIMUM RPM, THEN ENRICH SLIGHTLY
FLAPS	UP
RUNWAY	LEVEL, DRY, HARD SURFACE
WEIGHT	2150 LBS

TAKE OFF SPEEDS:

LIFT OFF	70 MPH/61 KTS
50 FT	80 MPH/70 KTS

WIND COMPONENT DOWN RUNWAY KNOTS	SEA LEVEL				2000 FT				4000 FT				6000 FT				8000 FT			
	OAT F	OAT C	GROUND ROLL FEET	TOTAL OVER 50 FT OBSTACLE FEET	OAT F	OAT C	GROUND ROLL FEET	TOTAL OVER 50 FT OBSTACLE FEET	OAT F	OAT C	GROUND ROLL FEET	TOTAL OVER 50 FT OBSTACLE FEET	OAT F	OAT C	GROUND ROLL FEET	TOTAL OVER 50 FT OBSTACLE FEET	OAT F	OAT C	GROUND ROLL FEET	TOTAL OVER 50 FT OBSTACLE FEET
0	23	-5	836	1331	16	0	953	1510	9	-13	1089	1717	2	-17	1247	1956	-6	-21	1430	2232
	41	5	930	1478	34	1	1062	1680	27	-3	1215	1913	20	-7	1393	2182	13	-11	1600	2495
	59	15	1030	1635	52	11	1178	1861	45	7	1350	2123	38	3	1550	2425	31	-1	1784	2777
	77	25	1137	1803	70	21	1303	2055	63	17	1495	2347	56	13	1719	2686	49	9	1981	3079
	95	35	1251	1982	88	31	1435	2262	81	27	1649	2586	74	23	1899	2963	67	19	2191	3402
15	23	-5	642	1195	16	0	738	1361	9	-13	850	1552	2	-17	980	1773	-6	-21	1132	2030
	41	5	717	1331	34	1	826	1517	27	-3	953	1733	20	-7	1100	1984	13	-11	1274	2274
	59	15	799	1476	52	11	921	1685	45	7	1063	1928	38	3	1230	2210	31	-1	1426	2538
	77	25	886	1631	70	21	1023	1865	63	17	1182	2136	56	13	1370	2452	49	9	1590	2820
	95	35	979	1796	88	31	1131	2058	81	27	1310	2359	74	23	1520	2711	67	19	1766	3122
30	23	-5	471	1084	16	0	548	1237	9	-13	638	1415	2	-17	743	1621	-6	-21	867	1860
	41	5	530	1209	34	1	617	1383	27	-3	720	1584	20	-7	840	1817	13	-11	982	2089
	59	15	594	1344	52	11	693	1539	45	7	808	1765	38	3	945	2028	31	-1	1105	2335
	77	25	663	1488	70	21	774	1706	63	17	904	1959	56	13	1057	2255	49	9	1239	2600
	95	35	738	1642	88	31	861	1895	81	27	1007	2167	74	23	1180	2498	67	19	1384	2884

Fig. 3-13. In order to meet charted takeoff performance, you must use the flaps as recommended by the manufacturer.

as "IAS at 50 feet" (Fig. 3-14). The flight-test examiner will look for it during your short-field demonstration.

TAKE-OFF DISTANCE — HARD SURFACE

ASSOCIATED CONDITIONS			TAKE-OFF SPEEDS	
POWER	FULL THROTTLE		LIFT OFF	70 MPH/61 KTS
MIXTURE	LEAN TO MAXIMUM RPM, THEN ENRICH SLIGHTLY		50 FT	80 MPH/70 KTS
FLAPS	UP			
RUNWAY	LEVEL, DRY, HARD SURFACE			
WEIGHT	2150 LBS			

WIND COMPONENT DOWN RUNWAY KNOTS	SEA LEVEL			2000 FT			4000 FT			6000 FT			8000 FT		
	OAT °F	°C	GROUND ROLL FEET	TOTAL OVER 50 FT FEET	OAT °F	°C	GROUND ROLL FEET	TOTAL OVER 50 FT FEET	OAT °F	°C	GROUND ROLL FEET	TOTAL OVER 50 FT FEET	OAT °F	°C	GROUND ROLL FEET
0	23	5	836	1331	16	9	953	1510	9	13	1089	1717	2	17	1247	1956	6	21	1430	2232
	41	5	930	1478	34	1	1062	1680	27	3	1215	1913	20	7	1393	2182	13	11	1600	2495
	59	15	1030	1635	52	11	1178	1861	45	7	1350	2123	38	3	1550	2425	31	1	1784	2777
	77	25	1133	1803	70	21	1303	2055	63	17	1495	2347	56	13	1719	2686	49	9	1981	3079
	95	35	1251	1982	88	31	1435	2262	81	27	1645	2586	74	23	1899	2963	67	19	2191	3402
15	23	5	642	1195	16	9	738	1361	9	13	850	1552	2	17	980	1773	6	21	1132	2030
	41	5	717	1331	34	1	826	1517	27	3	953	1730	20	7	1100	1984	13	11	1274	2274
	59	15	799	1476	52	11	921	1685	45	7	1063	1928	38	3	1230	2210	31	1	1426	2538
	77	25	886	1631	70	21	1023	1865	63	17	1182	2136	56	13	1370	2452	49	9	1600	2820
	95	35	979	1796	88	31	1131	2055	81	27	1310	2359	74	23	1520	2711	67	19	1766	3122
30	23	5	471	1084	16	9	548	1237	9	13	638	1415	2	17	743	1621	6	21	867	1860
	41	5	530	1209	34	1	617	1383	27	3	720	1584	20	7	840	1817	13	11	982	2089
	59	15	594	1344	52	11	693	1539	45	7	808	1765	38	3	945	2028	31	1	1105	2335
	77	25	663	1488	70	21	774	1706	63	17	904	1959	56	13	1057	2255	49	9	1239	2600
	95	35	738	1642	88	31	861	1885	81	27	1007	2163	74	23	1180	2498	67	19	1384	2884

Fig. 3-14. The airplane's best-angle-of-climb speed is often stated as "at 50 feet."

HEADWIND COMPONENT. A headwind provides two benefits to your takeoff. First, with a headwind (of 10 knots, for example), there is already a wind flowing over the wings before you even start rolling. Also, with that 10-knot headwind you are 10 knots closer to rotate speed even before you advance the throttle, which means the speed to which you must accelerate is 10 knots less; thus, a shorter takeoff run is needed. Second, a headwind following lift-off reduces your groundspeed, which gives you extra (possibly precious) seconds to gain extra yards of altitude as you climb toward the obstacle.

Wind rarely blows straight down the runway; it usually crosses at an angle. To apply the wind benefit to the takeoff chart, you must convert the existing wind direction and velocity to its direct headwind strength — called *headwind component.*

Make this headwind component conversion with three steps: *First,* look to the wind sock to determine the angle at which the wind is crossing the runway. Does it cross within 30 degrees of runway alignment? Within 30 to 60 degrees? Greater than 60 degrees? *Second,* determine the wind velocity. Wind socks are designed to stiffen at 15 knots. Thus, a one-third droop indicates 10 knots; a 45-degree droop, 8 knots; a two-thirds droop, 5 knots. *Third,* estimate the existing headwind component with a rule of thumb:

1. If the wind sock swings within 30 degrees of runway alignment, estimate the headwind to be the same as the wind velocity.

2. If the wind blows 30 to 60 degrees across the runway, estimate the headwind component as one-half the wind velocity.

3. Estimate the headwind as zero if the wind's angle to the runway exceeds 60 degrees.

RUNWAY SURFACE. Many aircraft manuals give numbers for only a dry, paved surface. Short fields, however, often have less than this ideal surface. In the absence of manual numbers for turf runways, figure that

1. If the grass is mowed and dry, increase the "paved" distance by 20%.

2. Make that fudge factor 30% if the grass is wet or needs mowing.

3. In the event of standing water or mud on an unpaved strip, treat the takeoff with a soft-field technique; the required distance can increase 50% or more.

Takeoff performance tables chart the seven variables fairly well. But even after carefully calculating the distance required, errors can occur. Many of your calculations are based on estimates and there are many opportunities for error—simple mistakes in arithmetic, a simple mis-reading of a chart value, a headwind that decides to die. For these reasons, you must apply two *fail-safe* measures when preparing for a short-field departure. The first measure is intended to prevent a chancy departure from even getting underway. The second is intended to stop the action once the roll begins, but before a hazardous situation arrives.

FAIL-SAFE MEASURES. When planning a short-field departure, incor-porate the fail-safe measure intended to prevent a possible dangerous situation from even occurring. Allow an adequate safe margin of run-way length—calculate total distance to clear an obstacle, and then add a 50% margin of safety. For example, consider conditions that require a calculated distance of 1800 feet. Unless you are a real ace, don't even try this takeoff unless the runway length exceeds 2700 feet (1800 feet × 150%).

Your second fail-safe, the measure that intends to stop the action, once under way, before it can turn to tragedy, calls for an abort point that allows safe stopping distance, as well as adequate climb distance to the obstacle. If you have a runway 150% of the required length, its mid-point should provide a safe margin.

You need to mark this abort point with a recognizable feature—the wind sock, a parked plane, a runway intersection. If there are no convenient landmarks, you may need to make your own; just taxi down the runway to the point that looks equidistant from both ends, get out of the plane, and mark the point with a piece of paper weighted down with a clod, or pin the paper to the turf with a pencil.

If you are not off the ground by the time you reach your abort marker, you can stop safely. Run on the ground beyond your marker, however, and you might be in a *must-do* situation. If you can't get aloft in time, you might hit the fence before you can get the plane stopped.

I am convinced that if pilots had made a timely abort, 90% of the takeoff accidents would never have occurred. Accordingly, I advise my students that when confronting a takeoff, keep in mind your *primary* plan—to abort. Then, if everything still looks okay during the run, put your *secondary* plan into play—to complete the takeoff.

When demonstrating your short-field procedure on the flight test, be sure to *tell* the examiner of your fail-safe measures. Examiners are *very* interested in your degree of safety consciousness, but they are not mind readers.

■ Short-Field Procedure

Once you have compared the performance charts to the conditions present for takeoff, you have half the equation in hand—the plane's ability to perform. The balance rests in the procedures you use to capitalize on the plane's ability.

Here are several tips you might find useful when demonstrating short-field departures on the flight test:

1. Verify engine performance before you take the runway. In performing a normal takeoff, glance at the tachometer early in the takeoff run. But short runways are often rough to the point where there is just too much noise, movement, and distraction to evaluate the tachometer or hear an engine misfire. Make a full-power runup before the takeoff roll begins. Look for the rpms that the manual states that a static runup should produce. Listen for suspicious engine roughness.

2. Conduct your short-field demonstration just as though you were facing the real thing. *Use* all the runway available to you. If the taxiway enters the runway several yards up from the threshold, back-taxi to that threshold.

3. Use the flap extension recommended by the manufacturer for short-field operations. Some manufacturers specify a short-field extension quite different from that used for normal departures.

4. Expend as little runway length as possible during the takeoff run. Four "don'ts" help deliver rapid acceleration:

a. Don't perform the run with underinflated tires. A tire gauge preflight check is in order.
b. Don't *jam* the throttle forward; a smooth push works fine. Overloading the carburetor with fuel causes a hesitation in engine power.
c. Don't let your toes rest on the brakes and don't try to steer with brakes — either move lengthens the ground run considerably.
d. Don't apply so much back pressure to the yoke during the run that you add aerodynamic drag. Use just enough pressure to let you *feel* the weight and friction of the nosewheel lessen.

5. Know the manufacturer's recommended rotate speed and apply positive rotation at that point. Rotate to the pitch attitude that you know (from practice) will result in the plane's stated best-angle-of-climb speed. Once at the altitude that clears the flight-test simulated obstacle, adjust pitch to achieve best-rate-of-climb airspeed.

6. As you climb through pattern altitude, retract the flaps one increment at a time. Readjust trim to maintain best-rate-of-climb speed as you fly toward cruise altitude. Then, give a glance to see if the examiner looks satisfied with your short-field demonstration.

■ Soft-Field Takeoff Execution

Those fields that turn to mush after a summer's rain present a very serious problem to most modern light aircraft. The planes of yesteryear, on the other hand, had no problem, for they were designed with soggy strips in mind — soft strips were common in those days.

Today's modern wings, props, and engines are more than adequate for soft fields; certainly more efficient than the old planes. But with modern planes, the underpinnings cause problems. Most light planes today are tricycle-gear equipped — the nosewheel carries the wings level, with the prop *very* close to the uneven surface to pick up damaging clods. And, that nosewheel wants to *plow* its way through the rough. Also, today's small, space-saving tires are just the right size to drop into every rut that they cross.

The high-stepping Wacos and Fairchilds of yesteryear, however, had tailwheels that held props high and away from damage (Fig. 3-15). Even at taxi speed, the uptilted wings provided lift to lessen the wheel load. And those big main tires could have been yanked from a Model T Ford.

Fig. 3-15. Light planes of yesteryear had tailwheels and oversized main tires that held props high above the turf.

They simply rolled across the ruts and mud with each rotation.

When confronting a soft field, then, you would be well advised to *simulate* tailwheel rigging with your tricycle-geared aircraft. It's easy to do.

Simulating Tailwheel Rigging

Begin to simulate a tailgear from the moment you begin to move after starting the engine. Conduct all of your ground travel with generous back pressure to the yoke. By doing so, you gain two advantages over the soft surface. First, like the tailwheeler, your plane's wings ride with an upward tilt. Even at taxi speed, this provides extra lift to ease the main tires across the soft spots.

Second, the back pressure lifts the nosewheel somewhat above the surface and, at the same time, raises the propeller arc for extra inches of safety. You have, in a sense, temporarily equipped your airplane with a tailwheel.

Soft-field takeoffs require some special techniques designed to meet the situation — techniques that you should make clear to the examiner in your flight-test simulation.

Soft-Field Techniques

PRETAKEOFF CHECK. If the infield or taxi areas are as soft as the runway, you may not want to stop once you get the plane rolling. Do *not* attempt, however, to perform your pretakeoff checks while taxiing. This practice invites an accident. Rather, perform your checks while still in the ramp area, before taxi begins.

Once taxi begins, use every reasonable effort to maintain a consistent taxi speed. Steer around the very softest spots, and if these spots are too big to avoid, add a bit of power before you reach them. Avoid taxi power in excess of 2000 rpm, however. Higher power often produces an erratic taxi, which can bounce a propeller tip into the dirt before you can react to stop it. When you're demonstrating your soft-field technique across a paved surface, *tell* the examiner the thoughts going through your head.

Once you reach the runway, keep your plane moving right into the takeoff run. And once you start the run, keep full-aft yoke to further lighten the nosewheel load as soon as possible. As the takeoff run speed increases, slightly decrease yoke pressure. The full yoke deflection, once speed increases, creates aerodynamic drag, which actually retards acceleration. Maintain a back pressure that lets you *feel* the nosewheel just skim the surface, for best overall acceleration. In short, you will have simulated tailwheel rigging with your tricycled airplane.

With the yoke held moderately aft, the plane will lift off before it reaches normal rotate speed. That's right — you get the plane into the air *before* it reaches flying speed because if left mushing through the mud, the plane may *never* accelerate to normal rotate speed. You can get away with this action due to a phenomenon called *ground effect*.

GROUND EFFECT. Wingtip vortices develop anytime that a wing produces lift (Fig. 3-16). These vortices are a principal source of aerodynamic drag. You can reduce much of this drag if you keep the vortices from forming. And, you can prevent the vortices from forming by flying in close proximity to the ground, which interferes with the circular motion of a developing vortex (Fig. 3-17). This in turn reduces aerodynamic drag. The interruption of vortex formation is called ground effect. The wings, while in ground effect, are highly efficient and the plane will fly at a velocity slightly lower than its normal lift-off speed.

There is, however, one serious drawback, and you must take a pre-

Fig. 3-16. Cyclonic wingtip vortices are a by-product of lift, and they create drag.

Fig. 3-17. The close proximity of the runway surface prevents the full development of wingtip vortices.

cautionary measure. In light aircraft, ground effect is effective only up to 12 feet, or so, off the ground (or about one-third of the wing span). This means that if you allow your plane to climb out of ground effect before you accelerate to normal flying speed, you can encounter a stall as vortices form to create normal aerodynamic drag. Therefore, as the wheels clear the runway, it is important to immediately lower the nose slightly to keep the plane skimming low over the runway until you accelerate to normal rotate speed. Once normal rotate speed is attained, pitch the nose for normal climbout.

The distance needed to clear a soft field is difficult to predict (it could easily exceed the manual's takeoff charted value by 50%). Yet, the takeoff chart provides valuable information. It helps you to put a soft-field fail-safe measure into play.

FAIL-SAFE MEASURE. Since the soft surface, excess back pressure, and delayed acceleration combine to cost unpredictable extra distance, you need a fail-safe measure. First, calculate the manual's total-to-clear-obstacle distance. Now, since much of the plane's acceleration occurs *after* the wheels clear, figure your required distance *from that point*.

Planning and executing a soft-field takeoff is a head game, as well as one of decisive, physical action. While your flight-test demonstration clearly shows your physical actions, you must *tell* the examiner how you thought through the procedure. Do all of this and your flight-test demonstration of soft-field procedures should pass with flying colors.

IN REVIEW

The FAA has pinpointed the takeoff as the time and place where most flying accidents occur.

Normal Takeoff and Climbout

Perform your takeoff in a sequence of seven specific steps:

1. Taking the runway
2. Making the takeoff run
3. Lifting off
4. Flying the initial climbout
5. Leaving the pattern
6. Climbing to cruise altitude
7. Leveling to cruise altitude and airspeed

Work to avoid any delay of action once the tower issues a "cleared for takeoff" clearance.

If departing a non-controlled airport, clear the area for traffic with a 360-degree clearing turn before taking the runway.

Remember, *all* runways are active at a non-controlled airport.

When taking the runway, position your plane directly astride the center stripe.

The flight examiner looks for four specific qualities in a takeoff run:

1. Most rapid acceleration to rotate speed
2. Verification of normal engine performance
3. Maintenance of positive directional control
4. Correction for any crosswind

A smooth throttle application delivers the best acceleration.

For maximum acceleration, avoid deflecting the control surfaces unnecessarily.

For best acceleration, avoid excess back pressure during the takeoff run.

Glance at the tachometer early in the takeoff run to confirm engine performance.

Avoid using differential braking as a means of directional control.

Anticipate the directional disturbances of *slipstream effect* and *adverse yaw* during the takeoff run.

Your best defense against a crosswind during the takeoff run is the positive use of *adverse yaw.*

The flight examiner expects to see two elements of control during lift-off:

1. Precise airspeed control
2. Positive directional control, which compensates for crosswind and P-factor

Precise adherence to the manufacturer's recommended rotate speed is essential to best takeoff performance.

■ Correct for a crosswind at lift-off by establishing a wind correction angle of about 1 degree for each knot of crosswind component.

Compensate for P-factor with right rudder pressure at lift-off.

■ Demonstrate four elements of precision flying as you fly your initial climbout:

1. Airspeed control
2. Directional control
3. Traffic avoidance
4. Correct pattern exit

Best-rate-of-climb speed is that airspeed that delivers the maximum gain in altitude for the *time* involved.

■ After rotation, establish the manufacturer's best-rate-of-climb speed and trim for that value.

Retract any extended flaps one increment at a time. Re-trim with each retraction to maintain the exact climb speed.

Torque and crosswind will disturb directional control. Compensate with right rudder and wind correction angle.

Most mid-air collisions occur at non-controlled airports during either the takeoff or landing phase of the flight.

To avoid traffic:

1. Spot other traffic

2. Alert traffic to your presence

Five steps of the climb to altitude:

1. Traffic awareness
2. Precise airspeed control
3. Precise heading control
4. Proper powerplant operation
5. Accurate level-off at cruise altitude

The engine cowl may restrict forward visibility when in the climb attitude.

Momentarily lowering the nose, or executing a shallow S-turn, will give a quick look ahead.

Five basic actions help alert other pilots of your presence:

1. Climb with your landing lights on
2. Waggle your wings periodically
3. Climb with your transponder on
4. Fly with your strobes burning
5. Give progressive position reports on the Common Traffic Advisory Frequency (CTAF) when departing a non-controlled airport

Maintain the manufacturer's recommended best-rate-of-climb speed throughout the climb. Use trim.

Engine torque will try to veer the plane to the left of course during the climb. Compensate with rudder trim or rudder pressure.

During the climb to altitude, fit your powerplant management efforts into three categories:

1. Monitoring engine gauges
2. Maintaining the manufacturer's recommended climb power setting
3. Managing the fuel flow

If you note an irregular engine instrument indication, advise the examiner and perform the appropriate discrepancy check.

Three common producers of power reduction during the climb:

1. Throttle creep
2. Carburetor ice
3. Improper fuel mixture

An extensive climb to altitude will require periodic adjustment to the fuel mixture control.

Allow yourself an unrushed level-off by leading the cruise altitude by 50 feet for each 500 feet per minute that you are climbing.

When leveling at cruise altitude, help the plane accelerate. Maintain climb power until you gain cruise speed.

During level-off and acceleration, vary rudder pressure to compensate for the changes in torque force.

Short-Field Takeoff and Departure

There are two basic elements you wish to display in your short-field demonstration:

1. An understanding of the airplane's takeoff performance chart that allows an evaluation of the takeoff environment at hand
2. Pilot skill that launches the plane with a minimum ground run and the most efficient initial climbout

Follow the aircraft's procedure manual to the letter when conducting a short-field takeoff. Manufacturers state their tested and proven procedures with the express purpose of keeping pilots within the planes' designed limits of operation. Take the plane beyond these limits and you suddenly become a test pilot.

■ The *operations section* of the plane's flight manual often recommends special techniques for short-field departures.

■ Proper aircraft loading (weight and balance) is critical to departure safety.

■ Takeoff and climb-performance charts calculate departure performance against seven environmental elements:

1. Obstacle clearance distance
2. Field elevation
3. Outside air temperature
4. Loaded weight of aircraft
5. Flap position and airspeed
6. Headwind component
7. Runway surface

The total distance to clear an obstacle is the value of interest to pilots, even if no obstacles are present.

Be prepared to modify the distance-to-clear value if the actual obstacle exceeds 50 feet.

Field elevation increases takeoff distance for three reasons:

1. Takeoff run increases
2. Rate of climb suffers
3. The airspeed indicator gives erroneous readings

High temperature has the same detrimental effect on takeoff performance as does high field elevation. Determine runway temperature from your plane's outside air temperature (OAT) gauge.

While most takeoff performance charts allow for loaded weight, few mention balance. Proper balance is assumed and must be in evidence.

Aircraft manuals clearly state the required flap position needed for safe short-field performance.

Best-angle-of-climb speed is that airspeed that delivers the maximum altitude in the least amount of distance flown.

Estimate your headwind component by the position of the wind sock.

In the absence of manual distance values for turf runways, apply some rules of thumb:

1. If the grass is mowed and dry, increase the "paved" distance by 20%.
2. If the grass is wet or needs mowing, make your fudge factor 30%.
3. If standing water exists, or the runway is muddy, treat the takeoff with a soft-field procedure. The distance required could easily exceed the extra 50% of the manual's stated value for a dry, paved surface.

Plan two fail-safe measures when facing a short-field departure:

1. Add a 50% margin of safety to your calculated distance required, to determine the minimum safe distance needed.
2. Plan an abort point that allows a safe margin in which to brake to a stop.

When demonstrating your short-field procedure, remember to *tell* the examiner your planned fail-safe measure.

Verify engine performance before taking the short runway, with a full-power runup.

Use *all* the runway available to you, even when demonstrating from a long runway. Back-taxi if necessary.

Use the manufacturer's recommended short-field flap position.

Soft-Field Takeoff

When confronting a soft field in tricycled aircraft, use the controls to simulate tailwheel rigging.
■ Use generous aft pressure on the yoke during ground travel on a soft surface.
Do *not* attempt to perform your preflight checks while taxiing; run the checks before taxi begins.
■ Once underway, plan ahead to maintain a consistent taxi speed.
■ Avoid taxi power in excess of 2000 rpm.
Once the takeoff run gets underway, begin decreasing the yoke to the point where the nosewheel skims the surface.
Soft-fields often require the pilot to lift off below normal rotate speed.
■ If lift-off below normal rotate speed is required and achieved, keep the plane flying within *ground effect* until the airspeed reaches normal rotate speed.
When departing a soft-field, plan an abort point that allows the climb distance (plus 50% margin of safety) required to clear any obstacle.

FLIGHT-TEST GUIDELINES

Takeoff and Departure

The flight-test examiner will ask you to demonstrate correct takeoff and departure procedures designed to meet the needs of three departure situations:

1. Normal takeoff and departure.
2. Short-field takeoff and departure.
3. Soft-field takeoff and departure.

Regardless of the demonstrated situation of the moment, there are a number of elements common to each that your examiner wants you to employ:

a. Referral and adherence to an adequate pretakeoff checklist.
b. Clear for traffic throughout the maneuver.
c. Position flaps in accordance with the manufacturer's recommended departure setting.

    d. Align the airplane astride the runway center line.

    e. Verify headwind and crosswind components.

    f. Begin the roll with full aileron into the crosswind and adjust the deflection during acceleration.

    g. Advance the throttle smoothly to maximum allowable takeoff power.

    h. Verify engine performance early in the takeoff roll.

    i. Maintain directional control astride the center line.

       1) Adequately correct for crosswind.

       2) Compensate for slipstream effect, engine torque, and P-factor.

    j. Maintain a straight departure course along an extended runway center line path until a pattern turn is required.

       1) Compensate for torque.

       2) Correct for wind drift.

    k. Retract flaps and gear at the appropriate time.

       1) Flaps—at safe altitude and airspeed.

       2) Gear—after a positive rate of climb is achieved, and after remaining runway length no longer allows a discretionary landing.

    l. Maintain takeoff power to a safe maneuvering altitude (pattern altitude).

    m. Complete an after-takeoff checklist.

    In addition to those elements common to *all* takeoffs and departures, the takeoff variants called for on the flight test ask for additional specifics.

■ Normal Takeoffs and Departures

1. Rotate at the manufacturer's recommended airspeed. (Advise the examiner of the rotate speed that you intend to use.)

2. Pitch to the attitude that delivers best-rate-of-climb speed (Vy) recommended by the manufacturer and maintain this airspeed ($+/-$ 5 knots) throughout the climb to altitude.

■ Short-Field Takeoff and Departure

1. Position the plane, prior to the roll, as close as practicable to the runway's threshold.
2. Adjust pitch as needed, to achieve the best acceleration.
3. Rotate at the recommended airspeed and pitch to the attitude that delivers best-angle-of-climb speed (Vx).
4. Maintain Vx (within 5 knots above, 0 knots below) until you clear the simulated obstacle (Note: Flight manuals for higher-performance aircraft may stipulate an airspeed other than Vx for obstacle clearance.)
5. After clearing the simulated obstacle, establish and continue the climb at Vy (within 5 knots above, 0 knots below).

■ Soft-Field Takeoff

When demonstrating your soft-field technique, the examiner will look for these actions:

1. Taxi onto the runway at a speed consistent with safety.
2. Taxi right onto the runway, align the plane for takeoff, and begin the takeoff run without interruption in forward motion.
3. Immediately into the roll, use back pressure to transfer much of the aircraft's weight from the landing gear to the wings.
4. Take advantage of ground effect to lift off as early as possible.
5. Accelerate to a safe flying speed while still within ground effect:

    a. If no obstacles, within 5 knots of Vy.
    b. If obstacles ahead, within 5 knots above best-angle-of-climb (Vx) speed, 0 knots below.
    c. If obstacle clearance is to be demonstrated, check the plane's flight manual for approved use of Vx. Be prepared to discuss the hazard of flying below Vx.

4.

Approaches and Landings

Students usually enjoy showing the examiner their landing expertise, and they usually do a good job. But, when you ask students what principal element of flying skill the examiner most wants to see, their answers often miss the mark when they answer, a smooth touchdown that "paints" the plane onto the runway, a landing right on the target, no crosswind drift. Of course, examiners want to see all of these elements, but they also want to see more—something even more basic—planning. The examiner knows that *all* the bits and pieces of skill for any maneuver come together only with planning.

And so it is with your landing demonstration. As you sit at the controls behind the rumbling engine on downwind, you must visualize the demonstration's five basic steps—each with its own goal—fitting precisely together, resulting in the perfect landing you are hoping for. The five basic steps to perform in sequence are (1) flying the traffic pattern, (2) descending to the runway, (3) rounding out, (4) touching down, and (5) rolling out. These five steps represent the degree of planning that the examiner hopes to see.

NORMAL APPROACH AND LANDING

■ Flying the Traffic Pattern

Why are traffic patterns rectangular? Why not a more efficient oval? The traffic patterns are rectangular simply because the square corners give a series of 90-degree clearing turns in a congested area. And just as important, perpendicular legs offer perpendicular movement between you and much of the traffic. This movement, perpendicular and at hard angles to one another, makes traffic easier to spot. (The hardest traffic of all to see is the plane dead ahead, traveling directly along your line of flight.) Therefore, use the pattern effectively. Search the perpendicular leg ahead of you before you get there. For example, on downwind search for traffic on *both* extended base legs. And upon turning base, search for traffic along the final legs of *your* runway, as well as all others. (At uncontrolled airports, especially, there is a need to search *all* final legs. At uncontrolled airports *all* runways are legally active.)

Early on the downwind leg, take a moment to pick some landmarks to help you visualize the downwind leg's ground track. Flying the traffic pattern is just another ground reference maneuver—first cousin to S-turns, turns around a point, and the rectangular pattern. And just as in these other ground reference maneuvers, it is very difficult to maintain a desired track unless you visualize the path on the ground, particularly if you're dealing with a wind.

Once you have landmarked your pattern, settle your plane to a pattern airspeed that slows the action, yet does not hold up traffic. Plan to use a pattern speed of 1.5 times stall (about 2200 rpm), but be ready and able to adjust that speed. If a traffic jam lies ahead, for example, don't hesitate to fly the entire pattern at approach speed; you have learned to fly slow.

On the other hand, if you realize that a jam is developing behind you, be ready to fly your entire downwind leg at near-cruise speed. Just remember to slow your plane to approach speed prior to turning final.

Before reaching the downwind point abeam to your touch-down area of the runway, quickly select three landmarks that will assist you in airspeed control, altitude management, and safety (Fig. 4-1).

1. Select a landmark to serve as a half-mile final-approach fix.
2. Pick a target on the runway to use as your touch-down bull's-eye.
3. Select a point on the runway to serve as your go-around point.

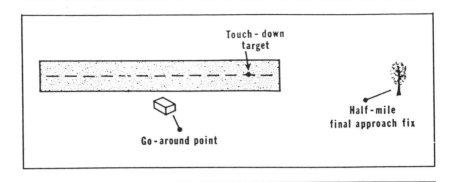

Fig. 4-1. Early on downwind, pick three landmarks that promote airspeed control, altitude management, and safety.

Old pros select and use these landmarks to deliver those seemingly effortless landings.

A half-mile final-fix point offers the advantage of providing an intermediate altitude objective during the landing descent. The pilot's objective is to position the plane 400 feet above this landmark at approach speed. (This 400 feet is appropriate for normal approaches. Situations such as very tall obstacles or the need to avoid the wake turbulence of a large plane may require a higher, steeper approach path.) A selected half-mile final-fix point on every landing also assures consistency. The pilot soon becomes an expert on that last half mile of the flight, for it has been flown at the same airspeed and altitude, landing after landing. Finally, a final-fix landmark is extremely valuable anytime the pilot needs to deviate from the standard landing pattern. Many times during your flying career, the tower will ask for your help with traffic spacing. For instance, you may be asked to widen your pattern, or make a downwind 360, or even turn to left base with a right 270-degree turn. The pilot with a final-fix landmark in sight is not bothered by these deviations, for that pilot knows that no matter what happens to the *rest* of the pattern, the final half mile will be just like every other final half mile flown before. Also, it is a simple truth that we all fly better when we know our exact destination, so give yourself the decided advantage of a visible landmark.

Select a highly visible landmark to serve as your touch-down target. Under normal circumstances, the second center line stripe from the runway numbers makes a good bull's-eye. It's far enough up the runway to prevent an undershoot, in which case you would catch your gear on the 8-inch threshold lip, yet lands you short enough to conserve runway length.

Finally, picking a go-around mark while still downwind is essential to safety. Don't put that critical go-around decision off until you are skimming above an evaporating runway with things starting to go wrong. If you *do,* you will find yourself committed to "putting it down," regardless of the risk. There is no doubt in my mind that most landing accidents would have never happened had the pilot made a timely abort. And this timely abort will occur only if the pilot has preplanned it.

Each pilot must decide when and how to apply flaps during the traffic pattern. Some pilots may elect to apply a third of their intended flap extension while still on downwind, a third on base, and the rest after turning final—and this is reasonable. Other pilots may prefer to extend half flaps on base, with full flaps applied on final—and this too is reasonable. The point is to develop your own procedure, use it consistently, and you soon become an expert with your flaps.

Some pilots feel that they must not use flaps when landing in a crosswind, for they are concerned that flaps might increase their plane's susceptibility to drift. But this is not the case. What *does* increase susceptibility to drift is groundspeed. The slower the groundspeed, the greater the wind drift (it is a function of time and force). Therefore, go ahead and use flaps when dealing with a crosswind. Just remember to fly your final approach at, or slightly below, your plane's proper no-flaps approach speed. By doing so you will be able to cope with the crosswind *while* benefiting from the advantages of flaps, which are (1) a reduced stall speed, (2) a steeper approach to conserve runway, (3) the ability to maintain the slipstream's *separation point* well aft on the wing's upper surface, and a (4) decreased pitch, thereby increasing forward visibility during the landing.

Think of flying your traffic pattern in terms of spotting traffic, visualizing the pattern's ground track, adjusting airspeed, selecting final-fix, touchdown, and go-around landmarks, preplanning a possible abort, and managing the flaps. Then, you are ready to tackle the descent to the runway.

■ Descending to the Runway

Ideally, your descent to the runway begins at pattern altitude abeam your selected touch-down target and ends on your short final over the runway with your plane at landing speed and at the altitude that ensures a bull's-eye touchdown. The three primary tasks during the descent from pattern altitude down to the runway are (1) managing a constant approach airspeed, (2) managing a consistent glide path, and (3) correcting for the crosswind.

Any good approach starts with your ability to control airspeed during the approach; this is so important. If you can hold a constant approach speed throughout the descent, then maintaining a consistent glide path is just a matter of making small power changes — 40 or 50 rpm of extra throttle if you slip below the glide path, or a bit less if the plane isn't descending steeply enough. *But,* if you fail to maintain airspeed, a smooth descent is extremely difficult. For then, you must simultaneously recapture the airspeed while you try to maintain a consistent glide path throughout two 90-degree changes in your flight path; too many variables present an almost impossible task.

A few tips to bring it all together: First, slow to approach speed while still on downwind, as you draw abeam your touch-down target. Decelerating this early in the approach slows the action, and you will have little trouble staying ahead of your plane. For the sake of standardizing your procedure, reduce throttle to an arbitrary setting suitable to your plane (about 1700 rpm is reasonable for most light trainers). You will probably need to adjust this initial setting during the descent, but *start* each approach with the same throttle and you soon will become expert at another element of the landing — setting up the approach. Apply carburetor heat at this time, if required.

Second, remember that proper trim is essential to good airspeed control. Readjust trim throughout the approach with each change in flaps or power.

Third, use your half-mile final-fix landmark to *guarantee* a smooth let-down from the downwind to the final. Start evaluating your rate of descent the moment you reduce throttle abeam your touch-down target. Plan a rate of let-down that places your plane at approach speed, 400 feet above your final-fix point.

Fly the descent smoothly, devoid of large power changes that would disturb your plane's pitch attitude, by reevaluating your descent path at every 50-foot increment on the altimeter. Simply *know* what you want the altimeter to read as you reach your final-fix point. Then as you complete your downwind and base legs, divide your attention between the landmark and your altimeter. The smooth let-down then becomes a simple matter of monitoring "distance to fly" with "altitude to lose," with very small (less than 50 rpm) throttle changes at each 50 feet if needed. Desperate changes of power are not necessary, and you stand an excellent chance of reaching your final-fix point at 400 feet AGL and approach speed. The old pros use this method, and there is no reason you shouldn't too.

Once over your final-fix point, start fine-tuning the let-down to your touch-down target. Chances are, the degree of headwind will call for some modifications in your final descent from 400 feet AGL. Use the *apparent motion* of your target to ensure a touchdown right on the mark. Here's how it works: If your target appears to move either upward or away from you, the glide path is too low and you will land short of the target; add a little throttle. If the target appears to move downward or toward you, you are overshooting; reduce throttle a bit. Once your target stabilizes with no apparent motion, your glide path is headed straight for it. (Your glide will flatten slightly in the last few feet, which will cause a touchdown slightly beyond the bull's-eye, but it will be close enough to pull an "ahh" from your flight examiner.)

Be mindful of the crosswind as you descend to the runway — and chances are, there *will* be a crosswind. Direct headwinds are rare, and a windless day is rarer still. As you fly your final approach, use both a crab and the cross-control capability of ailerons and rudder to correct for the wind.

As you initially roll from base to final, establish a crab angle to hold your flight path in alignment with the center line of the runway. A typical error when establishing this wind correction angle is one of *over-*correction. At typical light-plane approach speeds, it takes only about 1 degree of correction angle for each 1 knot of crosswind. Use a visual aid to fine-tune this correction. If the runway center line shows that a wind drift still exists after your initial crab angle, shift your heading 3 or 4 degrees to correct the problem.

You cannot continue to crab for the crosswind through to touchdown, for doing so would cause you to land at an angle to the runway that would impose a side load on the gear. The gear would probably not collapse, but the tires would screech — something the flight examiner does *not* want to hear (the examiner may ask: Was that a landing or a controlled crash?).

The examiner wants to see you shift from the crab to the cross-control slip as your final approach crosses the airport boundary with the touchdown near at hand. Many pilots have difficulty coordinating ailerons with rudder when using a slip to correct for a crosswind. They are bothered, not so much by the cross-control nature of the maneuver as they are by not knowing the true function of the ailerons and rudder. The true function of each is simplicity itself. Each control serves a separate task as the pilot corrects for the crosswind. Ailerons are used to prevent sidewise drift, and the rudder keeps the plane's fuselage aligned with the center stripe of the runway.

When you use ailerons to bank into the wind — say, to the right for right crosswind — the wing's lift (acting perpendicular to the wing's span) is deflected to the right. The force of this deflected lift tugs the plane against the wind. If the pilot applies insufficient right bank for the "tugging power" needed, the plane still drifts with the wind. If too much bank is applied, the plane actually moves sidewise into the wind.

To make a landing in a right crosswind, of course, if you bank to the right to prevent drift, the airplane's nose will also turn right unless you do something to prevent it. That's where rudder control comes in. Opposite rudder is simply used to prevent the airplane from turning toward the direction of the banked wings and to keep the fuselage aligned with the runway's center line.

You may think of the flight controls as performing two separate crosswind correction functions. Use ailerons for sidewise drift; use the rudder for runway alignment. But how much of each control should you use? The decision is easy, *if* you make use of one invaluable visual aid — the white center line on the runway. This visual aid gives you an instant-by-instant evaluation of your plane's wind drift and runway alignment. Again, picture the final moments of your approach with a right crosswind and with the center line in sight. If you start drifting to the left of the line, you need to increase aileron. If, on the other hand, you drift to the right (against the wind), you decrease aileron. And if you see your plane's nose cocked to the right of the center line, increase your "opposite rudder." If the nose points to the left, then decrease "opposite rudder." Use the runway center stripes to evaluate the magnitude of your aileron and rudder deflections — crosswind control on the final approach is as simple as that.

■ Rounding Out

The round-out segment of your landing sequence begins 30 or 40 feet above the runway with the upcoming touchdown still on target. During the brief moments of this segment, the three tasks the pilot must perform are (1) withdraw the power from the engine as the plane decelerates to touch-down speed, (2) change the plane's pitch attitude from that of approach to one of touchdown, and (3) maintain a precise crosswind correction.

Begin your round-out by slightly lifting the nose to break the approach airspeed, and by beginning to further reduce power as the runway threshold draws near. Keep your hand on the throttle during the round-out and coordinate your power reductions with your diminishing airspeed and increasing nose pitch attitude. Time your power reduction so that the throttle hits idle just as your tires touch. Carrying this small amount of power right through to touchdown guards against the prospect of "dropping it in," should you misjudge your round-out on the high side. Also, a few rpms carried to touchdown lets the engine respond quicker to a possible last-second go-around.

Use two visual aids to help change the plane's pitch attitude from one of approach to that of touchdown—the far end of the runway and the plane's nose. As you pass over the runway threshold, note the apparent space that lies between the tip of your plane's nose and the far end of the runway. Then, with each few feet of final descent to touchdown, raise the nose inch by inch to close the space between the nose and runway's end. For a perfect landing attitude in a modern tricycle-geared trainer, time your pitch changes so the nose just covers the far end of the runway as the tires touch.

Remain aware of the crosswind during the round-out. With touch-down imminent, it is time for precision correction. Remember: ailerons for drift, rudder for runway alignment. Don't count on a fixed control deflection to do the job. Wind is a fickle force—it rarely blows at a constant speed or direction, and you must stay nimble with ailerons and rudder. At this close point to landing, there is even little value in glancing at the wind sock. Several different puffs of wind normally exist between the sock and the touch-down zone where exactness counts. The runway center line, however, tells you everything you need to know about wind correction on an instant-to-instant basis. Also, remember that ground-speed affects the crosswind's force against the plane—the slower the plane moves, the greater the crosswind effect. During the round-out you are slowing from approach speed to touch-down speed, perhaps 20 knots of change. This alone calls for constant adjustments to control deflection. Again, the center line tells the story.

The round-out segment of the landing sequence is a coordinated blend of reducing power, increasing pitch attitude, and continued crosswind correction. It concludes with the engine at zero thrust, the nose in touch-down attitude, airspeed nearing stall, and the tires settling gently to the runway.

■ Touching Down

Treat your flight-test examiner to a good touchdown, which starts with the plane easing its main gear onto the runway and ends with the pilot gently lowering the nosewheel to the ground. The two primary tasks that are necessary during this brief segment of the landing sequence are (1) you must minimize the shock to the airframe, as a ton or so of airplane comes down on a chunk of pavement, and (2) you must maintain your crosswind correction.

You can best achieve a soft touchdown with a good follow-through.

A great deal of "stick travel" remains unused when the main wheels touch. A good follow-through uses this remaining yoke travel to ensure a super-soft landing. Just keep applying more and more back pressure to the yoke as the tires touch and the roll-out continues. Keep using all the back-travel that remains, to prevent the nosewheel from slamming into the runway. (This follow-through also prevents "stick freeze," which many pilots let happen in that moment before the tires touch. The wheels *will* hit with a hearty thump if you land with a motionless stick.)

Continue to use your ailerons and rudder to correct for the crosswind during the moments of touchdown. In this segment of the landing, your plane is part flying machine and part ground machine. The moments of transition, when the plane is least stable, are particularly susceptible to the wind. Most students know that they are supposed to land on the upwind tire when touching down in a crosswind. But, some feel that they must then level the wings to bring the raised downwind tire down to the runway. However, nothing could be further from the truth. You must keep aileron pressure into the wind as the upwind tire touches, then increase that pressure as the plane further decelerates. Don't worry about that raised tire. It comes down on its own accord when the time is right.

■ Rolling Out

Take care not to feel that your landing demonstration ends with a good touchdown. Keep your examiner smiling right through the roll-out. Here are a few tips:

1. Maintain directional control with rudder and nosewheel steering. For a smooth, non-swerving roll-out, avoid unnecessary braking until the plane slows to a fast-taxi speed.

2. Perfect directional control during the roll-out requires a 100% effort. Keep your eyes straight ahead, monitoring your direct path astride the runway center stripes. You cannot look down, for example, to retract the flaps, close the carburetor heat, or switch radio frequencies until the plane is fully stopped. To do so invites a swerve.

3. Remember that a plane rolling along the ground is subject to adverse yaw (aileron drag). Recall from the previous chapter that this means if you deflect ailerons, the plane tends to slew; that is if you turn the yoke to the left, the left aileron is raised and protected from the slipstream by the wing's curved upper surface. The right aileron, however, dips beneath the wing, digs into the slipstream, and the plane veers right. Therefore, unless you need control deflection with which to cor-

rect for a crosswind, it is best to let your ailerons streamline into the wind.

4. A crosswind during roll-out tends to weathervane your plane into the wind as that wind strikes the vertical stabilizer. During the roll-out segment of the landing, the examiner expects to see you compensate for any crosswind by making deliberate use of adverse yaw. If you experience a crosswind during your test landing, you can demonstrate your knowledge of how to use adverse yaw. However, the examiner may also ask you to explain the execution of this skill during the oral portion of the test. As an example, describe the situation of a crosswind blowing from the right—a wind that is pushing against the right side of your plane's vertical stabilizer, trying to weathervane the airplane to the right of center line. Your job is to keep the left aileron deflected downward into the slipstream with right yoke, so that it digs into the slipstream. As this left aileron digs into the wind, it counters the right-hand pull of the crosswind. As the plane slows—with drag diminishing—you maintain the left aileron's "digging power" by applying more and more control deflection. Perfect aileron movement during roll-out reaches full deflection just as the slowing plane reaches a stop.

5. Prevent a lurching stop at the conclusion of the landing the same way you do it in a car. In those last few yards of travel, begin easing brake pressure so that the plane comes to rest on its own, devoid of brake pressure.

So, if you perfect the five parts of the landing sequence, you just may elicit a "Not too bad" from your flight examiner.

SHORT-FIELD LANDING

I hope that you have the opportunity to land on some short fields. Of course, there was a time when turf short fields were the norm—not dangerously short, mind you, just short enough to allow a degree of uncertainty that required precise planning, flying, and a silent comment of *well done* at the end of the landing roll—an opportunity for self-challenge at the end of each flight.

In all probability, your first contact—on the flight test—with short-field flying, will occur during your oral questioning before you even get near the airplane. The examiner wants reassurance that (1) you know and understand the two primary goals of the short-field landing, (2) that you know and understand the significance of the eight variables that influence your plane's short-field landing performance and (3) that you know how to preplan fail-safe measures that reduce risk.

FLYING THE PRIVATE PILOT FLIGHT TEST

136

■ Short-Field Objectives

The short-field objectives are twofold in nature: your approach should be steeper than ordinary, and your touchdown should be at the slowest safe speed. These two objectives are meant to conserve runway length. To achieve these goals, you must know the eight variables that affect landing performance, and you must understand how to apply each one to the landing.

■ Landing Variables

Probably the best way to discuss the variables that affect landing distance is to relate them to the landing chart for the plane you fly. Most landing charts take each one into account (except the last one). So, with your plane's chart before you, refer to it throughout the discussion of each factor.

Runway Surface

The tabulation of landing distance is often predicated for a paved braking surface (Fig. 4-2). In most cases, there is a conversion factor given for the longer braking needs of the *unpaved* surface (Fig.4-3). If,

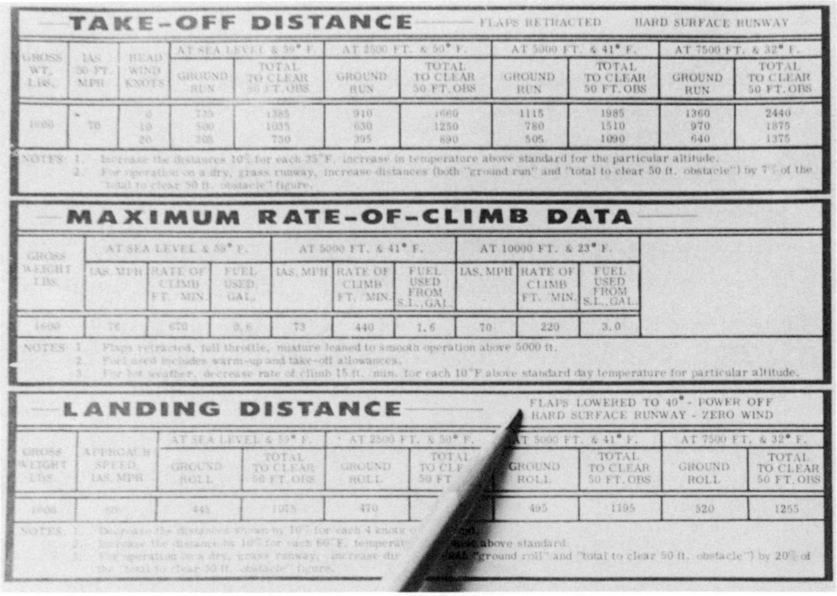

Fig. 4-2. Most landing performance charts relate their basic performance figures to a paved runway.

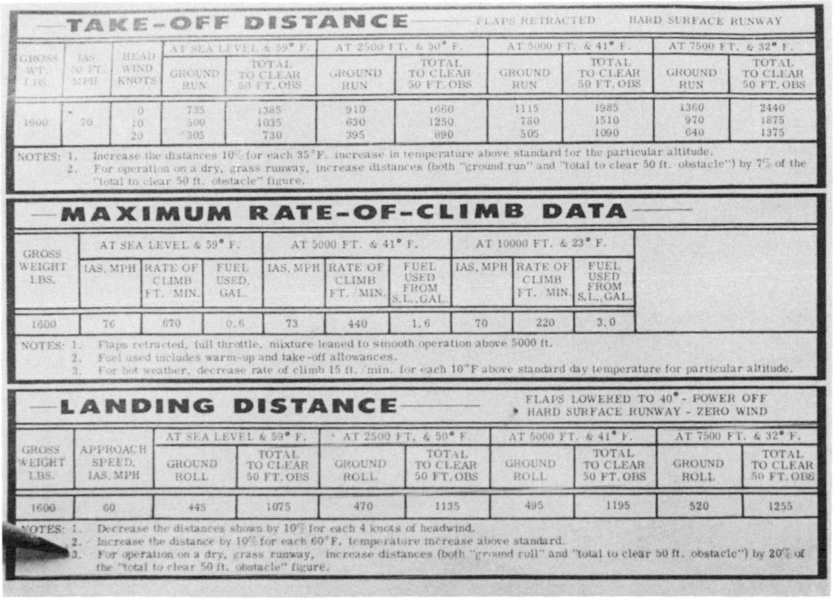

TAKE-OFF DISTANCE				FLAPS RETRACTED		HARD SURFACE RUNWAY		
GROSS WT. LBS.	IAS 50 FT. MPH	HEAD WIND KNOTS	AT SEA LEVEL & 59° F.		AT 2500 FT. & 50° F.		AT 5000 FT. & 41° F.		AT 7500 FT. & 32° F.	
			GROUND RUN	TOTAL TO CLEAR 50 FT. OBS	GROUND RUN	TOTAL TO CLEAR 50 FT. OBS	GROUND RUN	TOTAL TO CLEAR 50 FT. OBS	GROUND RUN	TOTAL TO CLEAR 50 FT. OBS
1600	76	0	735	1385	910	1660	1115	1985	1360	2440
		10	500	1035	630	1250	730	1510	970	1875
		20	305	730	395	890	505	1090	640	1375

NOTES: 1. Increase the distances 10% for each 35°F. increase in temperature above standard for the particular altitude.
2. For operation on a dry, grass runway, increase distances (both "ground run" and "total to clear 50 ft. obstacle") by 7% of the "total to clear 50 ft. obstacle" figure.

MAXIMUM RATE-OF-CLIMB DATA								
GROSS WEIGHT LBS.	AT SEA LEVEL & 59° F.			AT 5000 FT. & 41° F.			AT 10000 FT. & 23° F.		
	IAS, MPH	RATE OF CLIMB FT. MIN.	FUEL USED, GAL.	IAS, MPH	RATE OF CLIMB FT. MIN.	FUEL USED FROM S.L., GAL.	IAS, MPH	RATE OF CLIMB FT. MIN.	FUEL USED FROM S.L., GAL.
1600	76	670	0.6	73	440	1.6	70	220	3.0

NOTES: 1. Flaps retracted, full throttle, mixture leaned to smooth operation above 5000 ft.
2. Fuel used includes warm-up and take-off allowances.
3. For hot weather, decrease rate of climb 15 ft./min. for each 10°F above standard day temperature for particular altitude.

LANDING DISTANCE				FLAPS LOWERED TO 40° - POWER OFF		HARD SURFACE RUNWAY - ZERO WIND		
GROSS WEIGHT LBS.	APPROACH SPEED, IAS, MPH	AT SEA LEVEL & 59° F.		AT 2500 FT. & 50° F.		AT 5000 FT. & 41° F.		AT 7500 FT. & 32° F.	
		GROUND ROLL	TOTAL TO CLEAR 50 FT. OBS	GROUND ROLL	TOTAL TO CLEAR 50 FT. OBS	GROUND ROLL	TOTAL TO CLEAR 50 FT. OBS	GROUND ROLL	TOTAL TO CLEAR 50 FT. OBS
1600	60	445	1075	470	1135	495	1195	520	1255

NOTES: 1. Decrease the distances shown by 10% for each 4 knots of headwind.
2. Increase the distance by 10% for each 60°F. temperature increase above standard.
3. For operation on a dry, grass runway, increase distances (both "ground roll" and "total to clear 50 ft. obstacle") by 20% of the "total to clear 50 ft. obstacle" figure.

Fig. 4-3. Most landing performance charts offer a conversion factor for grass-field operations.

however, the unpaved performance is not mentioned, add 20% to the chart's paved distance for dry grass and 50% to the paved distance for wet grass.

Flaps and Airspeed

The stated landing distance usually assumes the use of full flaps (Fig. 4-4), which allows the pilot the slowest safe touch-down speed with which to minimize landing roll. To accomplish this, fully extend the flaps after you are well established on final and slow the plane to the lowest safe-approach speed recommended by the manufacturer. This slow speed delivers a relatively steep descent that consumes the least runway and is particularly useful when you are letting down over an obstacle.

Adherence to the recommended approach speed is critical to landing distance. An excess of only 5 knots can easily increase your landing distance by 20% in a typical light aircraft.

Field Elevation

Most light aircraft require about a 5% increase over sea-level landing distance for each 1000 feet of field elevation (Fig. 4-5). This increase in required landing distance occurs simply because you fly the approach

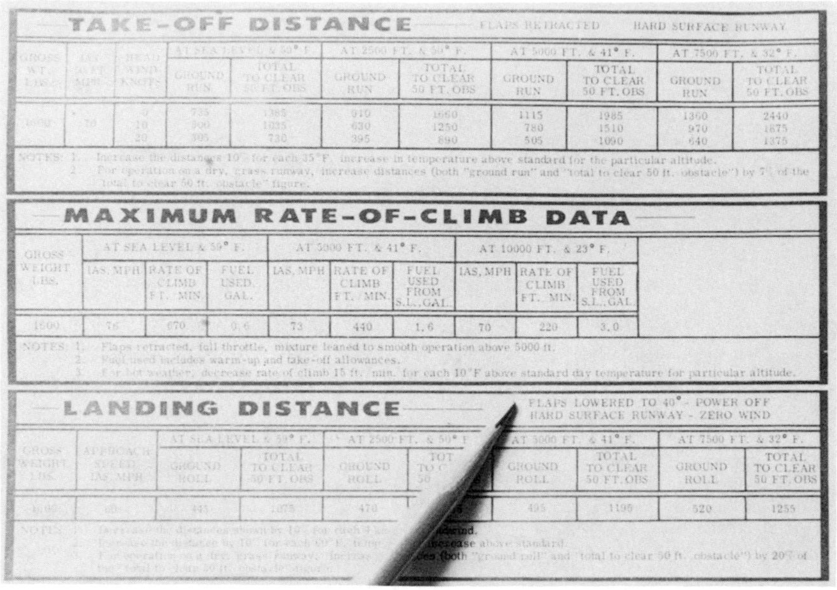

Fig. 4-4. It is essential to use the manufacturer's recommended flaps for the anticipated landing performance.

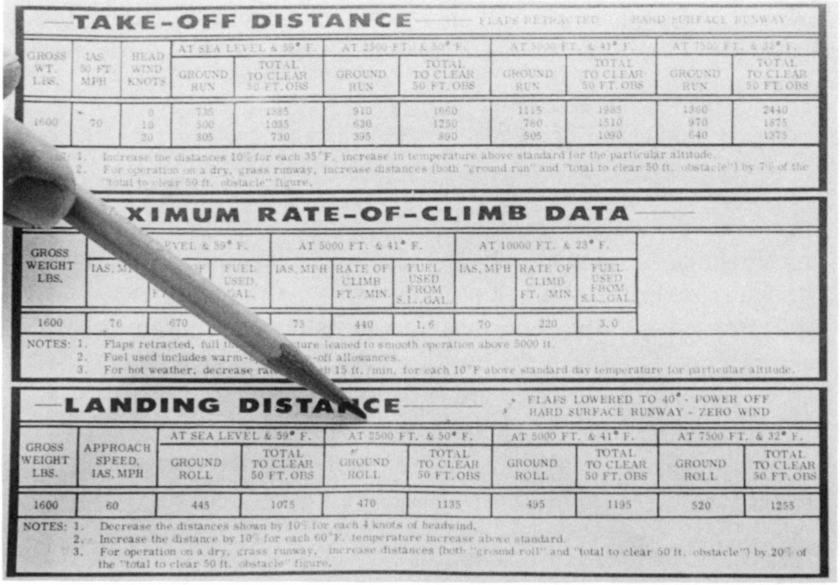

Fig. 4-5. Field elevation is a factor when predicting the landing distance required.

at the flight manual's recommended *indicated* airspeed, regardless of field elevation. Your *true* airspeed at these higher altitudes is, of course, slightly higher. This results in a higher groundspeed at touchdown and in turn produces a longer landing roll.

Aircraft Gross Weight

Some aircraft manuals provide expected landing distances for various aircraft loadings (Fig. 4-6); others provide the required distance for only the maximum load permitted. The heavier your plane is loaded, of course, the greater the distance required to bring the rolling momentum to a halt. This distance factor is significant. For lightly loaded small airplanes, the required stopping distance reduces about 5% for each 100 pounds under the plane's allowable gross weight.

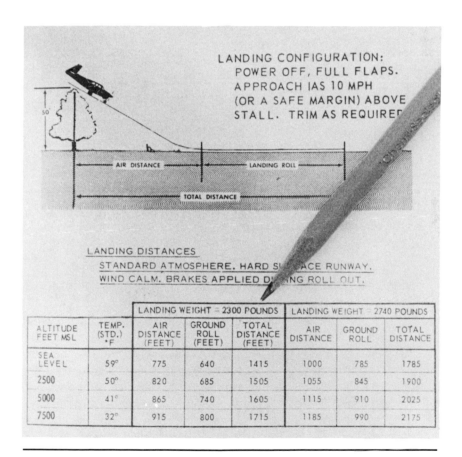

LANDING CONFIGURATION: POWER OFF, FULL FLAPS. APPROACH IAS 10 MPH (OR A SAFE MARGIN) ABOVE STALL. TRIM AS REQUIRED

AIR DISTANCE — LANDING ROLL

TOTAL DISTANCE

LANDING DISTANCES
STANDARD ATMOSPHERE. HARD SURFACE RUNWAY.
WIND CALM. BRAKES APPLIED DURING ROLL OUT.

ALTITUDE FEET MSL	TEMP. (STD.) °F	LANDING WEIGHT = 2300 POUNDS			LANDING WEIGHT = 2740 POUNDS		
		AIR DISTANCE (FEET)	GROUND ROLL (FEET)	TOTAL DISTANCE (FEET)	AIR DISTANCE	GROUND ROLL	TOTAL DISTANCE
SEA LEVEL	59°	775	640	1415	1000	785	1785
2500	50°	820	685	1505	1055	845	1900
5000	41°	865	740	1605	1115	910	2025
7500	32°	915	800	1715	1185	990	2175

Fig. 4-6. Typically, each 100 pounds of payload requires an approximate 5% increase in landing distance.

Headwind Component

The amount of headwind makes a big difference in landing distances. For example, the landing distance is usually shortened by about 20% with each 10 knots of headwind. Most landing performance charts estimate this reduced distance for you (Fig. 4-7).

TAKE-OFF DISTANCE — FLAPS RETRACTED   HARD SURFACE RUNWAY

GROSS WT. LBS.	IAS 50 FT. MPH	HEAD WIND KNOTS	AT SEA LEVEL & 59° F.		AT 2500 FT. & 50° F.		AT 5000 FT. & 41° F.		AT 7500 FT. & 32° F.	
			GROUND RUN	TOTAL TO CLEAR 50 FT. OBS	GROUND RUN	TOTAL TO CLEAR 50 FT. OBS	GROUND RUN	TOTAL TO CLEAR 50 FT. OBS	GROUND RUN	TOTAL TO CLEAR 50 FT. OBS
1600	70	0	735	1385	910	1660	1115	1985	1360	2440
		10	500	1035	630	1250	780	1510	970	1875
		20	305	730	395	890	505	1090	640	1375

NOTES: 1.   Increase the distances 10% for each 35° F. increase in temperature above standard for the particular altitude.
2.   For operation on a dry, grass runway, increase distances (both "ground run" and "total to clear 50 ft. obstacle") by 7% of the "total to clear 50 ft. obstacle" figure.

MAXIMUM RATE-OF-CLIMB DATA

GROSS WEIGHT LBS.	AT SEA LEVEL & 59° F.			AT 5000 FT. & 41° F.			AT 10000 FT. & 23° F.		
	IAS, MPH	RATE OF CLIMB FT./MIN.	FUEL USED, GAL.	IAS, MPH	RATE OF CLIMB FT./MIN.	FUEL USED FROM S.L.,GAL.	IAS, MPH	RATE OF CLIMB FT./MIN.	FUEL USED FROM S.L.,GAL.
1600	76	670	0.6	73	440	1.6	70	220	3.0

NOTES: 1.   Flaps retracted, full throttle, mixture leaned to smooth operation above 5000 ft.
2.   Fuel used includes warm-up and take-off allowances.
3.   For hot weather, decrease rate of climb 15 ft./min. for each 10° F above standard day temperature for particular altitude.

LANDING DISTANCE — FLAPS LOWERED TO 40° - POWER OFF   HARD SURFACE RUNWAY - ZERO WIND

GROSS WEIGHT LBS.	APPROACH SPEED, IAS, MPH	AT SEA LEVEL & 59° F.		AT 2500 FT. & 50° F.		AT 5000 FT. & 41° F.		AT 7500 FT. & 32° F.	
		GROUND ROLL	TOTAL TO CLEAR 50 FT. OBS	GROUND ROLL	TOTAL TO CLEAR 50 FT. OBS	GROUND ROLL	TOTAL TO CLEAR 50 FT. OBS	GROUND ROLL	TOTAL TO CLEAR 50 FT. OBS
1600	60	445	1075	470	1135	495	1195	520	1255

NOTES: 1.   Decrease the distances shown by 10% for each 4 knots of headwind.
2.   Increase the distance by 10% for each 60° F. temperature increase above standard.
3.   For operation on a dry, grass runway, increase distances (both "ground roll" and "total to clear 50 ft. obstacle") by 20% of the "total to clear 50 ft. obstacle" figure.

Fig. 4-7. A good headwind often is the determining factor when deciding to land on a short strip.

To estimate the headwind component on your flight test, look at the wind sock while on downwind. Wind socks are designed to stiffen at 15 knots, therefore, if the sock hangs at a 45-degree droop, you have a wind of 7 or 8 knots.

To convert the wind velocity to a headwind component, take note of the angle at which it swings across the runway. If the sock swings within 30 degrees of runway alignment, estimate the headwind component at full wind strength. If the sock blows 30–60 degrees across the runway, estimate the headwind component at half, and if the sock lies at a 60- to 90-degree angle to the runway, estimate the headwind component as zero.

Temperature

Warm temperatures, like field elevation, affect density altitude and adversely influence landing performance (Fig. 4-8); however, the influence of temperature is not as dramatic as is the influence of field elevation. Air thinned by heat (heat drives the air molecules apart) adds about 10% of the landing distance for each 40°F above the *standard temperature* for that field elevation. (Standard temperature is 59°F at sea level and decreases about 10°F for each 2500 feet of field elevation.) Prior to landing, you can estimate the runway temperature below by reference to your windscreen's outside air temperature gauge. Just add 4°F for each 1000 feet down to field elevation.

Fig. 4-8. Remember, summertime temperatures as well as an increase in field elevation can adversely affect landing performance. You need an additional 10% landing distance for each 40°F above the standard temperature for that field elevation.

Obstacle Clearance

Aircraft manuals chart two landing-distance values—one for the landing roll only, the other for total distance needed to clear a 50-foot obstacle, before braking to a full stop. It is this second value that is of prime interest to the pilot, for even without obstacles present you want to be at *least* that high as you cross the fence.

Remember, however, that obstacles rarely come in standard FAA 50-foot heights. So if your flight-test examiner stipulates a 100-foot obstacle, calculate the additional total distance required. First, figure the air distance needed for a 50-foot clearance (just subtract the *landing-roll* value from the *total-distance* value). Then, add this "per 50-foot" air travel distance for each additional unit of 50-foot obstacle-clearance distance needed.

Pilot Technique

The last element that affects the landing distance required doesn't show up on the manufacturer's performance chart. Each manufacturer predicates their aircraft's performance on an assumed basis of *average pilot ability.* I'm not exactly sure what *average* means, but I do know that things can go wrong. Sources of error are boundless: the headwind can die as you cross the threshold, you can touch down a trifle fast, brakes can fail, you can simply misread the performance chart. These and many other possible errors demand a margin of safety, so add a 50% margin for error to your calculated total distance required. For example, you calculate a needed distance of 1400 feet—don't even start the approach unless you have 2100 feet available (1400 × 1.5).

■ Short-Field Procedures

There is no doubt that a flight examiner considers the applicant's short-field demonstration as a showcase for planning and execution. If the maneuver is to be successful, the pilot must balance the factors of airspeed control, rate of descent, and power application while timing the approach within a proper traffic pattern. Therefore, carefully plan your procedure *and* make this planning evident to your examiner; don't hesitate to verbalize your plan as you execute the maneuver. Elaborate on the *whys* of what you are doing.

Size of Traffic Pattern

Many pilots erroneously scale down the size of their pattern when confronting a short runway. (I think they are just subconsciously trying to match a small pattern to a small runway.) However, you should use the same-sized pattern and same altitudes that you fly when landing at "Bigtown Municipal." Timing is all important when flying an approach to a short field, and an abbreviated pattern rushes your procedure to the point that the plane can quickly get ahead of your thinking.

Landmarking Your Approach

Take a brief moment while flying the downwind to select three landmarks to guide your descent and landing. Landmark your final-fix point, your touch-down target, and your missed-approach (go-around) point. On the final leg choose your final-fix point as a visible landmark that lies a half-mile from your touch-down target. (To quickly estimate a half-mile final, simply compare the distance to the known runway length. For example, a 2500-foot runway calls for a point about one runway length out.) Plan to cross your half-mile final-fix point at 400 feet AGL. This height ensures a fairly accurate descent slope that is neither dangerously low nor embarrassingly high. Then, small throttle adjustments should be all that is required to correct the descent for variations in normal headwinds. You avoid the need for a power-off, ton-of-bricks let-down or a high-power, plant-it-on-the-carrier-deck approach.

Every pilot needs a definite touch-down target to prevent either an undershoot or overshoot when landing on a short strip. Without such a visible landmark, you may not recognize a poor descent path until you are nearly down, and your last-second bid with the throttle is very apt to produce an uncontrolled, unsafe landing.

Your touch-down mark should be about a third of the way down the runway. This provides a safe margin against an undershoot, which might clip a fence, yet is short enough to conserve adequate stopping distance.

Finally, while still on downwind, select a landmark as your go-around point. The idea here, is to decide early on that you *will* go-around if your tires have not touched down prior to reaching that point, which still allows a safe stopping distance. If you delay this decision-making process until *after* you are confronted with a rapidly diminishing runway length, the anxiety of the moment may cause you to attempt an unsafe touchdown with insufficient runway remaining.

If you are sure that the strip provides 150% of the distance needed to land, then a go-around point at the runway's mid-point should provide a safe stopping margin.

■ Short-Field Technique

When landing at a short strip, begin your procedures while still on downwind. After picking your landmarks for the half-mile final fix, touch-down target, and go-around point, begin throttling back to approach speed. In single-engine light aircraft, a setting of 2000 rpm allows you to trim the plane to an initial approach speed of 1.3 × flaps-up stall speed while maintaining pattern altitude. You should plan to reach a downwind position abeam your touch-down target, in trim, with this approach speed well-in-hand, still holding pattern altitude. Slowing this early in the approach guarantees that the plane cannot get ahead of your thinking.

Immediately after drawing abeam the touch-down target, begin your descent by reducing throttle to 1700 rpm. As the plane begins to settle, adjust pitch attitude to maintain initial approach speed, and trim the plane to that speed at the reduced power setting. (Throughout the approach, *any* pilot must re-trim with each change in power or flaps, for precise airspeed control.)

Prior to turning base, gauge the remaining distance to reach your half-mile final-fix point and remember that your plan should be to cross that point at 400 feet AGL. Then, check the altimeter to see how much altitude you must lose to reach the 400 feet AGL. Then, throughout your glide to the final-fix point, re-evaluate your rate of descent at each 50-foot increment on the altimeter. Very small power changes (less than 100 rpm) will then allow a constant glide-slope correction if you judge your rate of descent to be too great or too small to achieve the desired height as you cross the final-fix landmark.

Begin your turn to base leg just after gaining a downwind position abeam the final-fix landmark. This assures an extra moment for planning as you turn final outside the final-fix point. While on base leg (still monitoring your descent by the altimeter and final-fix landmark), lower the flaps to the half-way point and re-trim for an unchanged airspeed in that configuration.

Extend full flaps immediately after turning final, while still outside the final-fix landmark. This provides a moment to slow the plane to the manufacturer's slowest, safe, full-flaps final approach speed (Fig. 4-9). Doing this while still outside the final-fix point allows time to adjust pitch and power to the full-flap landing configuration. In most light aircraft, the extension of full flaps changes the pitching moment and power needed. Depending on the pitching moment and drag produced by those full flaps, you may need to lower the nose slightly and add a tad more power to maintain the recommended final approach speed. You have time to do all this and recalculate and correct your rate of descent with small throttle movements, and still reach your goal of 400 feet AGL.

As you cross the final-fix at 400 feet AGL, in landing configuration and at the correct final approach speed, shift your attention to any obstacle that might be in your path. Your flight demonstration may

SHORT FIELD LANDINGS.

For a maximum performance short field landing in smooth air conditions, make an approach at 60 MPH with 40° flaps using enough power to control the glide path. After all approach obstacles are cleared, progressively reduce power and maintain 60 MPH by lowering the nose of the airplane. Touchdown should be made with power-off and on the main wheels first. Immediately after touchdown, lower the nose wheel and apply heavy braking as required. For maximum brake effectiveness, retract the flaps, hold full nose-up elevator, and apply maximum brake pressure without sliding the tires.

Slightly higher approach speeds should be used under turbulent air conditions.

CROSSWIND LANDINGS.

When landing in a strong crosswind, use the minimum flap setting required for the field length. Use a wing low, crab, or a combination method of drift correction and land in a nearly level attitude.

Excessive nose strut inflation can hinder nose wheel alignment with the airplane ground track in a drifting crosswind landing at touchdown and during ground roll. This can be counteracted by firmly lowering the nose wheel to the ground after initial contact. This action partially compresses the nose strut, permitting nose wheel swiveling and positive ground steering.

BALKED LANDING (GO-AROUND).

In a balked landing (go-around) climb, the wing flap setting should be reduced to 20° immediately after full power is applied. Upon reaching a safe airspeed, the flaps should be slowly retracted to the full up position.

In critical situations where undivided attention to the airplane is required, the 20° flap setting can be approximated by holding the flap switch for approximately two seconds. This technique will allow the pilot to obtain the 20° setting without having to divert his attention to the flap position indicator.

2-16

Fig. 4-9. Finding the manual's slowest, safe speed often requires a little searching through the pages.

include a *simulated* obstacle, its location and height stated by your examiner, or you may be confronted with an *actual* obstacle.

Simulated Obstacle

If your examiner stipulates a simulated obstacle for your short-field demonstration, *know* the landmark chosen for the obstacle, and its supposed height. If there is any doubt as to location and height, simply ask for clarification. With the final-fix landmark beneath you, focus your attention toward crossing the simulated obstacle—continue to evaluate your glide slope with each 50-foot decrease on the altimeter. Use small variations in power to place yourself over the obstacle with a safety margin of 50 feet higher than the obstacle.

Actual Obstacle

If there is an actual obstacle in your approach path, use a fail-safe visual aid that guarantees clearance. As you establish your final descent path over the final-fix point, take note of the "vertical gap" that appears between the tip of the obstacle and the touch-down target beyond (Fig. 4-10). As long as this vertical distance is increasing, you will clear the obstacle. If, on the other hand, that gap is diminishing, you may not have clearance.

After clearing the obstacle, concentrate on the touch-down target. Once past the obstacle, immediately start to slowly retard the throttle. This power reduction further steepens the last segment of your final

Fig. 4-10. The apparent gap between the tip of the obstacle and your touch-down target offers a visual aid to assure clearance.

approach, which conserves runway length. Plan this steady power reduction so that the throttle hits the stop just as the tires touch. (Avoid "chopping" power. If you chop power suddenly, you may let your airspeed get too low, and the sudden loss of power could allow a stall. Also, an engine at zero thrust does not respond quickly to sudden climb power, thus impairing a go-around, should one become necessary.)

Once you touch down on the short field, look straight ahead, for most short fields tend to be narrow—a look inside the cockpit to close the carburetor heat, for example, could cause a swerve and put you off the runway. Leave your flaps extended during the roll-out, for they provide aerodynamic braking. Also, hold the stick full aft as you apply heavy wheel-braking. This helps to prevent the nosewheel from slamming down as the brakes (aft of center of gravity) bite down. Then, just keep the nosewheel running along center line until the plane rolls to a stop.

Alternate Uses for Short-Field Technique

Is your short-field technique confined to landings on short runways? Definitely not. Use your short-field procedures anytime your *available* landing distance is limited by circumstances, regardless of overall runway length—these circumstances occur frequently in our day-in, day-out flying from long paved airports. An airliner, for instance, that lands on an intersecting runway reduces your usable runway length, since you must land short of that wake turbulence or touch down beyond the vortices. Similarly, the tower controllers impose a short field on you when they instruct you to land short of an intersecting runway. And, any pilot facing a rain-slick runway would be wise to use short-field techniques (without heavy braking). The same procedures go for a pilot who catches a whiff of hydraulic fluid on downwind and suspects faulty brakes.

Be ready to discuss these alternative uses for the technique as you and your flight examiner discuss short-field procedures.

SOFT-FIELD LANDINGS

Alas, those landing fields that turn to mush during the muddy spring are becoming harder and harder to find with each passing year. So many are giving up their ghostly echoes of the chug-chug-chugging Kinner and Curtiss engines to the concrete of condos and shopping malls.

This is a shame, and I urge you to seek out these fields before they totally disappear. Landing at any one of these soggy fields will give you two pleasures. First, we never seem to outgrow the childish delight of splashing about in a mud puddle. And when this splashing about is performed within a cockpit behind a rumbling engine, the delight is magnified tenfold.

Second, landing at one of these fields will put you in direct touch with our civil aviation history. I have noticed that most students and pilots who excel at the controls have in common an awareness of flying's "roots." And there is no finer place to find these roots than at these old flying fields.

You know, the history of civil aviation really isn't that old — it began to develop in the 1920s, following World War I. Some of those early planes are still flying, and you might expect to see, perhaps, a big, silver, 1937 Spartan Executive with its bulbous cowl filled with a ring of over-sized Wasp cylinders, in a hangar at one of the old, soft fields. Or maybe a saucy 1931 Monocoupe that, modified, ran in the golden age of air racing. Or even more likely, that army brat of World War II, the Stearman biplane of the early '40s. (Pilots who fly the close-coupled, geared Stearman offer a piece of landing advice: Touch down with the left brake locked, then you *know* which way its gonna ground loop.) Many of yesterday's planes are within those hangars; Fairchilds, Ryans, Wacos, and others. Each waiting for you to see, touch, and contemplate.

■ Soft-Field Landing Variables and Pilot Technique

You can best understand soft-field landings by studying the construction features of the old airplanes and the concerns of the pilots who flew them. Those planes of yesteryear exhibited four attributes that favored soft-field landings: (1) a low sink rate (extensive wing area) that prevented a hard contact with soft ground; (2) a slow forward speed (pronounced wing curvature) that defied a nose-over after touchdown; (3) a nose-high ground-handling attitude (tailwheel rigging) that held the prop away from damage; (4) a gear (oversized wheels) that could ride out the soft surface.

Let's discuss each of the four elements in turn.

Low Rate of Sink/Extensive Wing Area

The older planes had "acres" of wing area, which did not allow for much speed, but it sure created a slow rate of descent. And, why did the pilots treasure a parachute-like rate of sink? Simply because it helped prevent driving the descending gear into the soft surface.

The modern Skyhawk has a wing area of 175 square feet, which produces a wing loading of 14 pounds per square foot and a sink rate of about 750 feet per minute. By comparison, the 1928 Curtiss Robin (Fig. 4-11) had a wing area of 225 square feet, giving it a wing loading of 10 pounds per square foot and a sink rate of about 480 feet per minute. This significantly lower rate of descent greatly favored soft-field landings.

Fig. 4-11. The Curtiss Robin of 1928 was designed when soft fields were the norm.

Your technique can help you attain a comparable sink rate in the Skyhawk. Remember how you touched down with the power at zero thrust for a short-field landing. Once over the obstacle at 1700 rpm, you slowly retarded the throttle with a movement that had it reach the stop just as your tires touched down. Well, when approaching a soft field, leave the throttle at 1700 rpm right through the approach. Then, as you cross the fence, make a small power reduction to 1500 rpm, which allows an increase in touch-down pitch attitude. Finally, leave the throttle at 1500 rpm through touchdown. (Of course, you may need to make small adjustments to this power setting to hit your touch-down target; however, plan to land with about 1500 rpm.) Carrying this power through touchdown reduces the typical rate of sink by 200 feet per minute. This may not match the Robin's sink rate, but it comes close.

Slow Forward Speed/Pronounced Wing Curvature

A look at the cross section of the wing of a Skyhawk and a Robin shows a significant design difference. The wing of the Robin is thick, with a highly curved upper surface. While preventing a fast cruising speed, this extreme curvature allows a very slow landing speed of 37 knots, even before the era of flaps. The slow touchdown, when the wheels met the soft, clinging surface, was followed by a slow forward motion that produced little pitching moment that might stand the plane on its nose. However, even a tricycled plane can flip with an incorrect soft-field touchdown.

The cross section of the Skyhawk's wing, on the other hand, is slender — a compromise between a slow landing speed and a fast cruising speed. The clean-winged Skyhawk touches down at 53 knots.

A combination of features and technique make soft-field landings possible. The modern planes have flaps. A full-flap approach at the manufacturer's slowest recommended safe speed is the answer to the soft field. This slow approach speed is often stated in the aircraft manual, under either short-field or soft-field operations (Fig. 4-12). If this slowest safe speed is not stated, then calculate the value yourself. First, determine the plane's level-wing, power off, wheels down, full-flap stalling speed (Fig. 4-13). Multiply this speed by 1.3 for the slowest safe approach speed. (Note: This formula applies only to single-engine, unmodified, standard light aircraft, as do *all* procedures in this book.)

A word of caution: The slow speed has you flying on the "razor edge" of the plane's limits, so do not chop power. If you do and you lose a few knots of airspeed, you can easily stall. Also, if a go-around becomes necessary, make sure you build up good flying speed before you start retracting the full flaps, otherwise you might find yourself trying to keep flying when below the no-flaps stalling speed. Remember that FAA statistic: Most accidental stalls occur during the approach to landing when low altitude puts you at a disadvantage for a successful recovery. It *can* happen the moment you lose sight of the potential danger.

Nose-High Ground-Handling Attitude/Tailwheel Rigging

The old planes like the Robin were tailwheeled. The tailwheel gave them the two important advantages of holding the spinning prop high above the runway and tilting the wings upward to "grab" a few angles of attack. Even at taxi speed, this helped lift the plane across the mud and ruts.

Even with today's modern tricycled planes, you have to keep the prop away from the turf when rolling across soft, often rough, terrain. To do this, as soon as you touch down, first keep the yoke pulled full aft until you bring the plane to a stop. As with the older planes, this slow rolling speed, aft yoke, gives you a few extra degrees of pitch, which holds the prop higher. (This is comparable to "grabbing" a few angles of

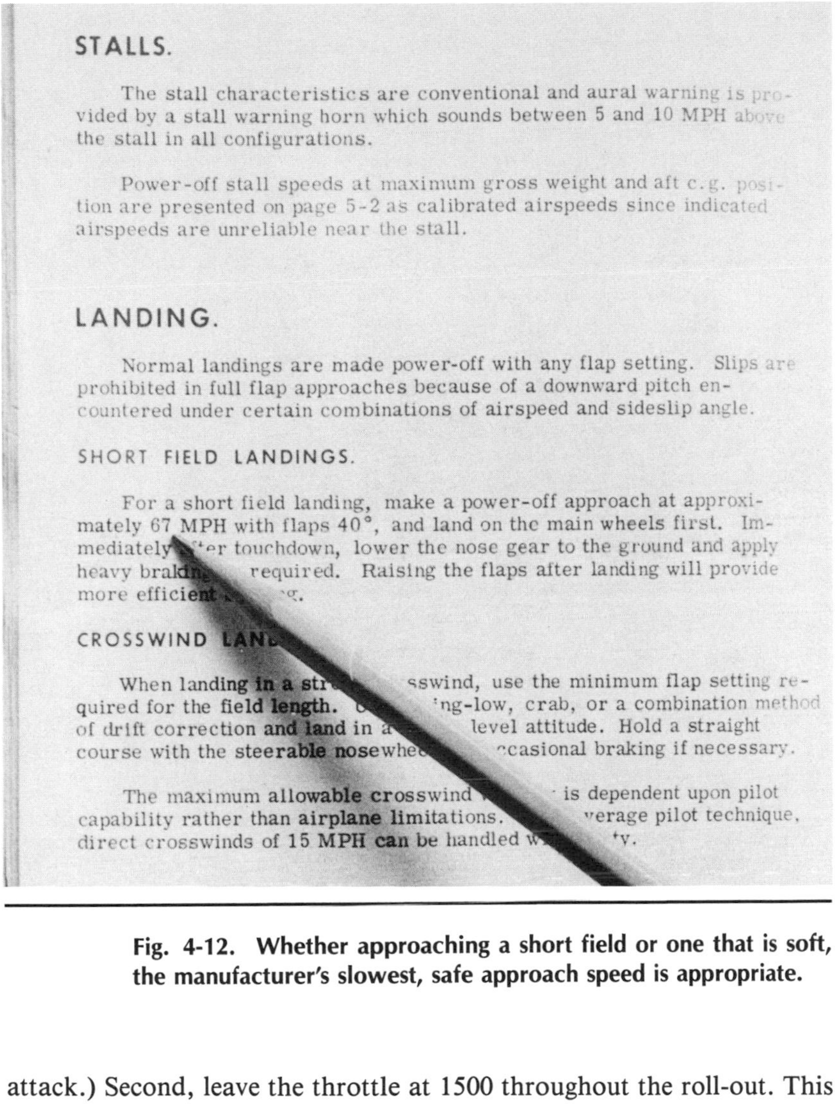

STALLS.

The stall characteristics are conventional and aural warning is provided by a stall warning horn which sounds between 5 and 10 MPH above the stall in all configurations.

Power-off stall speeds at maximum gross weight and aft c.g. position are presented on page 5-2 as calibrated airspeeds since indicated airspeeds are unreliable near the stall.

LANDING.

Normal landings are made power-off with any flap setting. Slips are prohibited in full flap approaches because of a downward pitch encountered under certain combinations of airspeed and sideslip angle.

SHORT FIELD LANDINGS.

For a short field landing, make a power-off approach at approximately 67 MPH with flaps 40°, and land on the main wheels first. Immediately after touchdown, lower the nose gear to the ground and apply heavy braking as required. Raising the flaps after landing will provide more efficient braking.

CROSSWIND LANDINGS.

When landing in a strong crosswind, use the minimum flap setting required for the field length. Use wing-low, crab, or a combination method of drift correction and land in a near level attitude. Hold a straight course with the steerable nosewheel and occasional braking if necessary.

The maximum allowable crosswind velocity is dependent upon pilot capability rather than airplane limitations. Using average pilot technique, direct crosswinds of 15 MPH can be handled with safety.

Fig. 4-12. Whether approaching a short field or one that is soft, the manufacturer's slowest, safe approach speed is appropriate.

attack.) Second, leave the throttle at 1500 throughout the roll-out. This delivers some prop blast across the upturned elevator, which pushes the tail down to tailwheel height. (Don't worry about a too-fast rolling speed—the soft surface prevents that.)

Riding Out the Soft Surface/Over-Sized Wheels

Those over-sized tires on the older planes were the perfect ally for the soft-field pilot. They let the plane just ride *over* the ruts and mud.

By contrast, the modern Skyhawk gear has small tires. These wheels and tires are designed to reduce weight and drag in favor of a faster

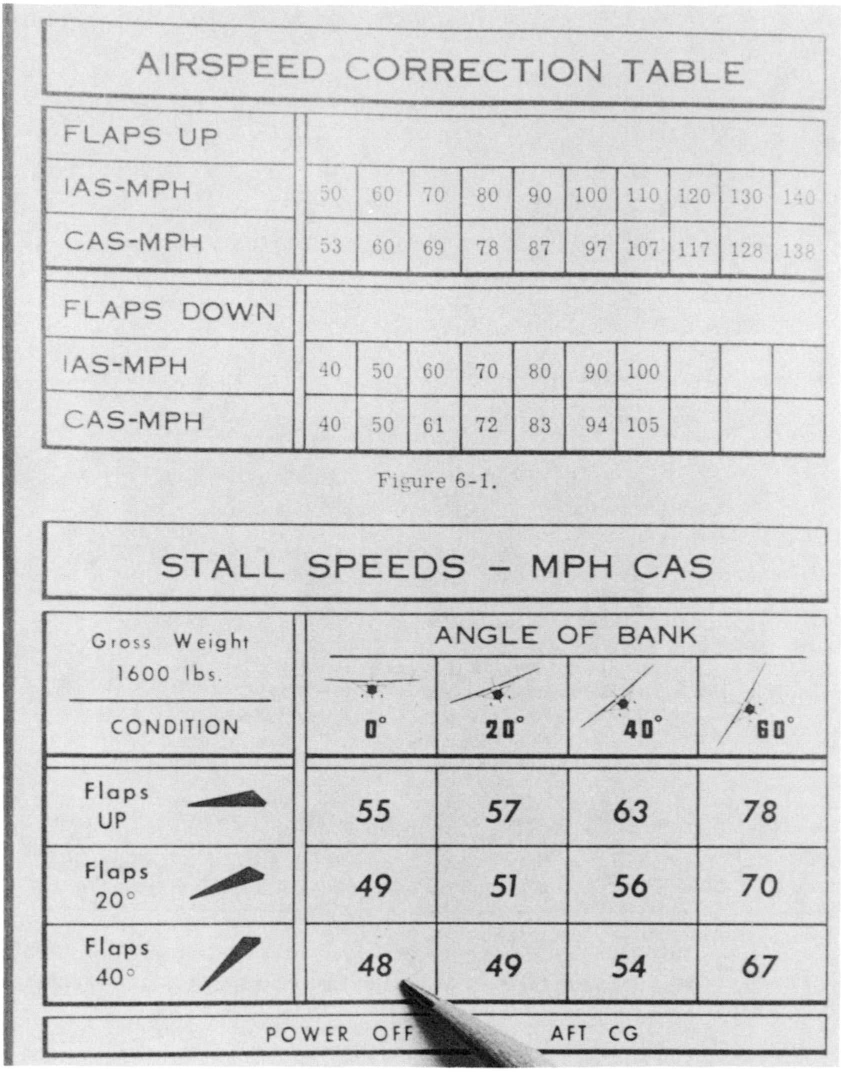

AIRSPEED CORRECTION TABLE

FLAPS UP										
IAS-MPH	50	60	70	80	90	100	110	120	130	140
CAS-MPH	53	60	69	78	87	97	107	117	128	138
FLAPS DOWN										
IAS-MPH	40	50	60	70	80	90	100			
CAS-MPH	40	50	61	72	83	94	105			

Figure 6-1.

STALL SPEEDS — MPH CAS

Gross Weight 1600 lbs. CONDITION	ANGLE OF BANK			
	0°	20°	40°	60°
Flaps UP	55	57	63	78
Flaps 20°	49	51	56	70
Flaps 40°	48	49	54	67

POWER OFF	AFT CG

Fig. 4-13. If the manufacturer's slowest, safe speed is unavailable, calculate your safe speed: 1.3 × landing configuration stall speed.

cruise speed—great for enroute but not too good for soft-field landings. You may best achieve soft-ground handling with smaller tires by leaving the flaps fully extended until you shut down in the ramp area.

Combining these four attributes with piloting skill, you *can* meet the challenge of a soft-field landing. As you perform the maneuver, do not hesitate to verbalize your understanding of these attributes to the flight-test examiner; your soft-field demonstration should work out just fine. I'm not worried.

◼ SLIPS TO A LANDING

In the past, pilots slipped to a landing to shorten their approach when clearing an obstacle into a short strip. When slipping, they flew down the final *sidewise* to mash the side of the plane through the air. The plane sank like a horseshoe. Today, you don't really need to slip for a steep descent because flaps do the same job, only better. In a sense the maneuver of slipping is becoming a "lost art" of flying. Why then, are you asked to learn slips and to demonstrate them on the flight test?

First, you learn them because they are so much fun to do, and pilots seem to get great pleasure from slipping their planes down the final leg. Having fun is a big part of flying.

Second, the slip to a landing is an excellent flight-test showcase of piloting skill. The maneuver demonstrates your ability to maneuver accurately while the flight controls are in a "cross-control" situation (ailerons banked in one direction, rudder turned in the other). Show the examiner your best stuff.

◼ Slip Technique

To understand how a slip works, follow me through an approach. Imagine I've just turned from base leg and am established on final. At this point I realize my approach is too high and decide to steepen it with a slip. My first move is simple — close the throttle. It doesn't make much sense to try and slip away altitude while maintaining a power setting that tends to reduce my rate of sink. (Also, avoid slipping with flaps extended, for flaps tend to reduce rudder control in a slip.)

Once the throttle is closed, I remind myself of the crosswind's direction across the runway. Although the direction of the crosswind has little effect on the slip during the let-down, sometime prior to touchdown I will need to shift from the "slip-to-a-landing" to the "slip-for-crosswind correction." This correction will require that I touch down with wings banked into the crosswind, and this shift is easier if I already have my "down wing" into the wind.

Assume that I'm facing a landing with a left crosswind. I establish the slip by using pronounced aileron to bank the wings left, while simultaneously pushing firmly on the right rudder to yaw the plane to the right of the forward flight path. With equally firm aileron and rudder pressures, the plane descends straight ahead, but flies sidewise through the air—banked left, yawed right.

This side-saddle flight attitude creates excess drag, which steepens the approach considerably. Yet, I can easily control this rapid sink rate. If I want to steepen the descent even further, I add equal rudder and aileron pressures. The extra rudder swings even more of the plane into the wind, while the extra aileron keeps the plane descending straight ahead. If, on the other hand, I need to shallow the descent, reducing control pressures does the job. With a little dexterity I can control the let-down to a touchdown right on target.

The trick is to maintain the proper approach speed throughout the slip. Doing so isn't easy, for the airspeed indicator may give erroneous readings during the slip. This inaccuracy stems from a pitot tube that is no longer aligned directly into the wind and a static port that may no longer have neutral air pressure flowing across it. Add to this the fact that a plane's airspeed wants to change considerably when I deflect the fuselage against the wind.

The best way to maintain the proper airspeed throughout the slip is to know my airplane well enough to recognize the *sound* of a proper airspeed's slipstream across the windscreen, and to recognize the *feel* of the controls that a proper approach airspeed produces. I must rely heavily on my ears and my fingers to tell me how high or how low I must hold the nose to maintain approach speed. Then, at the conclusion of my descent, I must remind myself to return the nose to its normal approach attitude as I roll out of the slip in preparation for landing. If I fail to make this adjustment, the airspeed will go awry just as the runway comes up to meet me.

Once I have brought the plane down accurately toward the touch-down target, I must shift from "slip-to-a-landing" to "slip-for-crosswind correction." To do this I simply release enough rudder pressure to let the fuselage align with the flight path and runway center line, and maintain just enough aileron to correct for the crosswind.

The only difference between a "slip-to-landing" and a "slip-for-crosswind correction" lies in the degree of the slip; the degree of control pressures. When you lose altitude with a slip to a landing, you use equally strong rudder and aileron pressures. You deflect the alignment of the fuselage well to one side of the flight path with strong rudder while banking in the opposite direction—equally strong opposite aileron keeps the plane's flight path on course. The wings are banked left, for example, while the nose is deflected to the right. Conversely, when a slip is used for crosswind correction, only enough aileron is used to prevent drift, while only enough rudder is used to keep the fuselage *aligned* with the flight path. In correcting for a left crosswind, for instance, the wings are banked left, but the nose points straight ahead.

Enjoy your demonstration during the flight test. Do the job well and both you and your flight examiner will appreciate your skill for the lost art of flying.

IN REVIEW

Plan to perform your flight-test landing demonstrations as five-step maneuvers:

1. Flying the traffic pattern
2. Descending to the runway
3. Rounding out
4. Touching down
5. Rolling out

Normal Approach and Landing

■ Use 90-degree turns to clear for traffic on the pattern leg that lies ahead.

Pick landmarks to help you visualize the downwind leg of the traffic pattern; use this visualization to prevent drift.

■ Plan to use a downwind pattern speed of 1.5 × stall.

While on downwind, select three landmarks that afford accuracy and safety to your landing:

1. A half-mile final-fix point
2. A touch-down target on the runway
3. A go-around point

■ A half-mile final-fix point offers an intermediate let-down goal.

The second center line stripe offers a safe touch-down target that conserves runway length.

Commitment to a highly visible go-around point is necessary to a safe landing.

Plan how you want to apply flaps during the descent to the runway.

There are three primary tasks to accomplish during the descent to the runway:

1. Maintain a constant approach speed

2. Manage a consistent glide path
3. Correct for the crosswind within the approach

Slowing to approach speed while still on downwind slows the action and prevents the plane from getting ahead of you.

Proper trim is the key to good airspeed control; re-trim with each change in flaps or power.

Use your half-mile final-fix point to assure a consistent let-down; plan to cross the fix at 400 feet AGL.

Use the apparent motion of your touch-down target to land right on the mark.

Use a combination of crab and slip to correct for the crosswind as you descend down final.

Use the runway center line as a gauge for appropriate aileron and rudder pressures.

The pilot's task during round-out is threefold:

1. Withdraw power as the plane decelerates to touch-down speed.
2. Lift the plane's pitch attitude from that of approach to one of touchdown.
3. Maintain a precise crosswind correction.

Time your power reduction during round-out so that the throttle hits idle just as the tires touch.

During round-out, time your pitch changes so the nose just covers the runway's far end as the tires touch.

When correcting for the crosswind during round-out, remember: ailerons for drift, rudder for runway alignment.

You have two primary tasks to perform during the brief moment of touchdown:

1. Minimize the shock to the airframe.
2. Maintain a crosswind correction.

Avoid the shock of touchdown with a follow-through that applies more and more back pressure to the yoke as the tires touch.

Keep applying aileron against the crosswind throughout the touchdown and into the roll-out.

During the roll-out, maintain directional control with rudder and nosewheel steering. Avoid the unnecessary use of differential braking.

Directional control during the roll-out demands 100% attention. Keep your eyes straight ahead.

■ A rolling plane's directional control is subject to adverse yaw (aileron drag).

■ During roll-out, your best defense against crosswind force is the positive use of adverse yaw.

■ Prevent a lurching stop at the conclusion of the landing roll. Begin easing brake pressure during those last few yards of travel.

■ Short-Field Landings

■ The examiner looks for reassurance in three areas of short-field know-how:

1. That you know and understand the two primary goals of short-field technique.
2. That you know and understand the seven environmental variables that influence a plane's short-field performance.
3. That you know how to plan fail-safe measures that reduce risk.

■ Two objectives of a short-field approach:

1. A steep approach that conserves runway length.
2. A slow touch-down speed that reduces landing roll.

■ There are seven basic environmental variables that affect landing distance:

1. Runway surface. (Turf surfaces increase the needed landing distance by about 20%.)
2. Flaps and airspeed. (Full flaps mean a landing-distance difference of about 30%. Each excess of 5 knots of speed increases the distance about 20%.)
3. Field elevation. (Each 1000 feet of elevation above sea level adds about 5% to the landing distance.)
4. Aircraft gross weight. (Each 100 pounds means a difference in landing distance of about 5%.)
5. Headwind component. (Each 10 knots of headwind reduces the landing distance by about 20%.)
6. Temperature. (Each 40°F above standard temperature adds about 10% landing distance.)
7. Obstacle clearance. (Remember, the charts calculate only a 50-foot obstacle.)

■ Plan fail-safe measures into your soft-field technique:

1. Adhere to a prechosen abort point.
2. Require a runway length with an adequate safety margin of charted distance × 150%.

■ When challenging a short field, you should use the same-sized pattern and altitudes that you fly when landing at Bigtown Municipal.

■ Soft-Field Landings

■ When confronting a soft field use a technique that:

1. Delivers a minimum sink rate, to avoid driving the main gear into the soft surface.
2. Produces a slow touch-down speed that helps avoid a nose-over.
3. Maintains a nose-high ground travel that keeps the prop clear of the turf.
4. Minimizes the plane's weight on the wheels rolling across a soft surface.

■ A low rate of sink and a soft touchdown may be achieved by carrying the descent power right through touchdown.
■ A low touch-down speed can be achieved with full flaps and the manufacturer's slowest safe recommended approach speed.
■ Nose-high ground travel may best be achieved with full-aft yoke and taxiing with reasonable power.
■ Taxiing with a nose-high attitude increases the wings' angle of attack to help lift the wheels across a soft surface.

■ Slips to a Landing

■ Slips are fun to perform.
■ Perform your slip with no flaps and power off.
■ Slip with the down-wing into the crosswind.
■ Equal aileron and opposite-rudder pressures guide the plane for a straight-ahead descent.
■ Presenting the side of the fuselage to the wind produces a steep approach.
■ Changing the control pressures equally regulates the rate of sink.
■ Expect inaccurate airspeed indications during the slip.
■ Slipping with the down-wing into the crosswind facilitates crosswind correction during the round-out and touchdown.

FLIGHT-TEST GUIDELINES

■ Approaches and Landings

The flight examiner will ask you to demonstrate techniques designed to meet the needs of four landing situations:

1. Normal landings
2. Short-field landings
3. Soft-field landings
4. Slips to a landing.

Regardless of the situation of the moment, there are a number of techniques common to each landing that you should use:

1. Refer and adhere to an adequate pre-landing checklist.
2. Clear for traffic with each leg of the landing pattern.
3. Use flaps in accordance with the manufacturer's recommendations.
4. Know and use the flight manual's landing performance charts.
5. Fly appropriate pattern altitudes and correct for wind drift within the traffic pattern.
6. Correctly manage power, flaps, and airspeed to deliver a constant descent.
7. Touch down astride the runway center line, with no crosswind drift.
8. Correct for crosswind during the roll-out astride the center line, concluding with smooth braking.
9. Refer and adhere to an appropriate after-landing checklist.
10. Each approach must show planning in the event of a go-around from a rejected landing:

    a. Explain the elements of a safe go-around (flaps/airspeed).
    b. Ability to execute a timely go-around.
    c. Ability to establish correct climb speed and trim.
    d. Ability to track straight ahead.

■ Normal Landings

1. Exhibit and be prepared to discuss the elements of a normal landing.
2. Establish and maintain the manufacturer's recommended approach airspeed ($+/-$ 5 knots).
3. Maintain a runway center line ground track on final.
4. Execute a smooth transition from approach altitude and power to the landing attitude and power.

5. Touch down within 500 feet upwind from a specified landing target.
6. Maintain positive directional control during roll-out.

■ Short-Field Landings

1. Discuss the elements present in a short-field attempt.
2. Adequately clear a simulated obstacle.
3. Establish a rate of descent that conserves runway length.
4. Establish and maintain the manufacturer's short-field flap configuration and approach speed down the final ($+/-$ 5 knots).
5. Touch down within 200 feet upwind from a specified landing target.
6. Apply appropriate braking to conserve runway length.

■ Soft-Field Landings

1. Explain the elements and technique of a soft-field landing.
2. Establish the manufacturer's recommended soft-field flap configuration.
3. Maintain the manufacturer's slowest, safe approach speed down final ($+/-$ 5 knots).
4. Touch down at a minimum forward speed with minimum rate of sink.
5. Use power and flight controls to assist plane across a simulated soft surface, during roll-out and taxi.

■ Slips to a Landing

1. Explain the purpose, technique, aircraft limitations, and airspeed indicator error when slipping to a landing.
2. Begin the slip from a point that allows a safe touchdown (400 AGL over the half-mile final-fix point).
3. Maintain a runway center line ground track.
4. Maintain the manufacturer's recommended approach speed through the slip.
5. Prevent excessive speed from developing as you roll out of the slip. Keep the center line "captured."
6. Touch down within 500 feet upwind from a specified touch-down target.

■ Landing Demonstrations

As your flight-test examiner requests demonstrations of each landing variation, perform your maneuvers as a sequence of logical steps.

Normal Landings

DOWNWIND LEG

1. Clear for traffic on the airport surface, ahead and to either side, and for planes entering or leaving pattern.
2. Reduce power to establish airspeed of 1.5 × stall speed (about 2200 rpm).
3. Establish correction for wind drift.
4. Refer and adhere to a written pre-landing checklist.
5. Select three landmarks with which to assist the approach and landing:

   a. Half-mile final-fix point.
   b. Touch-down target.
   c. Go-around point.

6. Reduce power to begin the descent (about 1700 rpm) when abeam the touch-down target.
7. Establish and trim to the recommended approach speed.
8. Plan a descent that will place you 400 feet AGL over the final-fix point.
9. Evaluate your rate of descent — distance to the final-fix versus altitude yet to lose. Evaluate glide path at each 50-foot increment on the altimeter. Use small power adjustments to maintain a consistent glide path.
10. Clear for traffic ahead and to either side.
11. Begin your turn to base with the final-fix point (ideally) slightly behind your shoulder. (Be prepared, however, to adjust your pattern to fit the traffic.)

BASE LEG

1. Roll out on a base-leg heading that corrects for wind drift.
2. Clear for traffic.
3. Lower the flaps one-half the setting you intend to use on landing and re-trim to maintain the proper approach speed. (Throughout your approach, re-trim with each change in flaps or throttle.)
4. Continue to monitor your rate of descent — distance remaining toward the final-fix versus 400 feet AGL at that point.
5. Clear for traffic ahead and to either side.
6. Begin your turn to final. If you have a trailing wind on base, begin this turn early to allow for drift in the turn.

FINAL LEG

1. Turn to final a little above 400 feet AGL, with the final-fix still slightly ahead.
2. Clear for traffic ahead, to either side, and on the ground.
3. Roll to a final-leg heading that establishes a wind correction angle.
4. Extend the flaps to final setting as you cross the final-fix point. Adjust pitch to maintain approach speed and re-trim.
5. Fine-tune your wind correction angle against an extension of the runway center line.
6. Begin a further smooth power reduction that leads your glide path directly toward the touch-down target.
7. Use the "apparent motion" of the touch-down target to gauge your glide path. Use small power adjustments accordingly.

    a. If the target is moving upward or away, you are undershooting—add a bit of power.
    b. If the target is moving downward or closer, you are overshooting—retard power a bit.
    c. If the target remains motionless—you are right on the mark.

ACROSS THE FENCE

1. Begin slowing to touch-down speed and attitude at about 50 feet AGL. Use a slow, smooth power reduction with a constant, smooth, slight back pressure on the yoke.
2. Use small power reductions and pitch changes that will result in zero thrust and touch-down attitude just as the tires touch the ground.
3. Shift your crosswind correction from that of a crab to one using slip.
4. Let the runway center line guide your controls to slip against the crosswind.

    a. Use ailerons (bank) for left or right correction to land astride center line.
    b. Use rudder (yaw) to keep the fuselage aligned with the center line.

DOWN THE RUNWAY

1. Touch down with an attitude that has the nose just covering the end of the runway.

    a. Smoothly apply remaining yoke back-travel to minimize the nose-wheel's shock against the runway.
    b. Keep the ailerons turned into the crosswind.

2. Perfect directional control during the roll-out demands 100% of your attention. Keep your eyes straight ahead.
3. Use rudder and nosewheel steering for directional control; avoid differential braking.
4. After rolling speed decreases, use smooth, steady braking to stop.
5. Exit the runway and refer to a written after-landing checklist.

Short-Field Landings

PREFLIGHT PLANNING

1. Explain to the examiner how you determine the landing distance required, using the plane's landing performance chart.
2. Show the examiner that you can use the *Airport/Facilities Directory* and weather information to determine the expected landing environment at the short-field destination:

   a. Field elevation.
   b. Runway heading, length, surface, known obstacles.
   c. Expected temperature at ETA.
   d. Expected surface wind (combined with runway headings to determine expected headwind and crosswind components).

SHORT-FIELD TRAFFIC PATTERN

1. Fly the short-field traffic pattern in accordance with *normal landing* procedures.
2. Do not "scale down" the pattern to "fit" the short runway.
3. Adhere to a written before-landing checklist.
4. Remember to clear for traffic with each turn in the pattern.

SHORT-FIELD APPROACH

1. Cross the half-mile final-fix point at 400 feet AGL at the manufacturer's recommended, slowest, safe airspeed, with flaps fully extended.
2. After crossing the final fix, use incremental power changes to cross the simulated obstacle with a 50-foot margin for safety.
3. After crossing the simulated obstacle, smoothly throttle back any remaining power to achieve a steep, power-off descent.

SHORT-FIELD TOUCHDOWN

1. Keep your eyes straight ahead — most short runways are narrow.
2. Keep the yoke full aft to prevent heavy braking from slamming the nosewheel into the ground.
3. Apply smooth but positive braking to shorten the landing roll.

Soft-Field Landings

DOWNWIND LEG

1. Clear for traffic on the ground, ahead and to either side, and for planes entering or leaving pattern.
2. Reduce power to a pattern speed of 1.5 × stall speed (about 2200 rpm).
3. Establish correction angle for wind drift.
4. Adhere to a written pre-landing checklist.
5. Select three landmarks with which to assist the approach and landing:

   a. Half-mile final-fix point.
   b. Touch-down target.
   c. Go-around point.

6. Reduce throttle to begin descent (about 1700 rpm), when abeam the touch-down target.
7. Trim to the recommended approach speed.
8. Plan a descent that will place you 400 feet AGL over the final-fix point.
9. Re-evaluate your rate of sink with each 50 feet of altitude lost.
10. Clear for traffic and begin your turn to base when slightly beyond a position abeam the final-fix point.

BASE LEG

1. Roll out on a base-leg heading that corrects for wind drift.
2. Clear for traffic.
3. Lower the flaps one-half the setting you intend to use on landing and re-trim to maintain approach speed.
4. Clear for traffic ahead and to either side.
5. Begin your turn to final.

FINAL LEG

1. Turn to final a little above 400 feet AGL with the final-fix still slightly ahead.
2. Clear for traffic ahead, to either side, and on the ground.
3. Roll to a final-leg heading that establishes a wind correction angle.
4. Extend the flaps to final setting and trim to the recommended slowest, safe approach speed to produce the slowest forward motion at touchdown.
5. Upon crossing the final-fix point at 400 feet AGL, *leave* the throttle

positioned at descent power (about 1700 rpm) to produce a low sink rate at touchdown.

6. Use small power variations and the apparent motion of the touchdown target to gauge your final descent path.

DOWN THE RUNWAY

1. Touch down while maintaining power (about 1500 rpm). Use full-aft yoke pressure to keep the nose pointing high.
2. Continue to apply full-yoke pressure throughout the roll-out; continue to provide power to the engine.
3. Conclude your landing by retarding the throttle and brake to a smooth stop.

Slips to a Landing

1. Begin your slip to a landing 400 feet AGL above the half-mile final-fix point.
2. Initiate your slip by correcting for wind drift with a crab angle.
3. Make the transition from the crab to the slip.

   a. Lower the upwind wing while you simultaneously apply enough opposite rudder pressure to hold course as you present the side of the fuselage to the slipstream.
   b. Quickly adjust pitch to the sound of the wind and the feel of the controls to maintain the correct approach speed.
   c. Gauge the rate of your descent using the apparent motion of the touch-down target on the runway. Adjust the magnitude of your slip to maintain an accurate glide path.
   d. Across the fence, shift from slipping to a landing to a crosswind correction. Use rudder pressure to align the fuselage with the center stripe. Use ailerons to correct side-wise wind drift.
   e. Touch down and roll out with standard crosswind control.

After you have earned your pilot certificate, take your plane to a country strip and have fun delivering perfect slip after perfect slip—enjoy reliving a lost art of flying.

5.

Emergency Procedures

Your flight examiner wants to see evidence of a constant awareness of potential emergency situations, and that you have a plan of action with which to meet them. As evidence of this awareness, your examiner may set the scene: You are flying some friends to a distant eatery that promises the best hamburgers and hot, cinnamony, deep-dish apple pie in the state. Blue skies—a few puffballs of fair-weather cumulus—a smooth roaring Lycoming—just enough ripple in the air to let you know you are flying—pleasant conversation. Then, the examiner may call for your response to a sudden, simulated emergency: Coughs and spits from beneath the cowl send a rumbling shudder of alarm through the aluminum fuselage. These simulated situations will fall into either of two general categories, equipment malfunctions or emergency landings.

EQUIPMENT MALFUNCTIONS

Before discussing specific malfunctions, let's list the few things you should do *anytime* a situation of concern comes your way:

1. If conditions and time permit, climb to a higher altitude. Not only will you increase your communication range, but with many potential problems extra altitude is like money in the bank.

2. Estimate the time you can remain aloft — remaining fuel is a factor, as is the nature of the malfunction. (A rough engine will usually get you home, for example, but rapidly falling oil pressure is another matter.) Plan to have your "bird at roost" before dark.

3. Communicate with an FAA facility — any facility will do the job. If you are already working a frequency, then stay on it. If you are unsure of a working frequency, go to the emergency 121.5, that's what it's there for. *All* FAA facilities monitor it.

Remember, you don't need to declare an emergency to use 121.5. Quite the contrary. It is important that you alert your ground support system *before* your situation becomes an emergency. In fact, once a situation deteriorates into a real emergency, your options are usually so limited that advice and support from the ground is of minimal value. In general, you should put out a call of concern any time (1) your fuel supply first comes into question, (2) you become unsure of your position, (3) an equipment malfunction arises for which you cannot effect an immediate fix, or (4) weather starts to look a bit scary — certainly before clouds force you to within 2000 feet AGL, visibility drops to 5 miles, or it looks like you may be forced to enter precipitation.

You do not need to declare an emergency to get help. Just tell someone that you have a concern and ask them to stick with you. They will.

When you ask for assistance, there are a few standard requests that you should make immediately:

    a. A radar or direction finding (DF) confirmation of your present position.

    b. The heading and estimated time enroute to an appropriate alternate airport.

    c. Advice in selecting an *appropriate* airport. If your concern is a misfiring engine, for example, you will want to land at an airport that offers maintenance facilities. Or, if your problem suggests a hazardous landing (like a gear-up), you will want an airport with fire and rescue equipment.

4. If finding the remedy to a mechanical problem confounds you, put in a call to a nearby FBO on their *Common Traffic Advisory Frequency* (unicom). Ask to speak to their mechanic.

5. Finally, make sure your passengers know the situation. The last thing a pilot wants is panicky passengers. When given the facts, however, most passengers will cooperate with a degree of calm. Just tell them of the problem, and reassure them that ground assistance has been alerted

and that you already have plans laid for a safe arrival, and that you can cope.

■ Rough-Running Engine

The most common equipment problem that a pilot may expect to encounter is a rough-running engine. This rarely develops into a serious problem, but the examiner will likely ask to see your response to the simulated situation.

The most common causes for a misfiring engine are a loss of fuel flow, carburetor ice, or improper ignition. The examiner expects you to use a checklist:

1. Mixture control — full rich.
2. Auxiliary fuel pump — on.
3. Fuel selector — switch tanks.
4. Throttle — restore cruise power.
5. Carburetor heat — full hot.
6. Magnetos — check each, return to "Both."
7. Use the smoothest throttle, mixture, ignition.

■ Unlatched Door in Flight

A door open in flight is no real problem — just a lot of wind noise — with most single-engine light aircraft. Positive control remains with the door open about two inches. If the door pops open on takeoff with ample runway ahead, stop and close it. If safe stopping distance is not available, continue the takeoff, level at 2000 feet AGL, and then close the door. The point is, don't wrestle with the door while you are flying close to the ground.

■ Overheating Engine

Engine overheating in flight is most often caused by restricted airflow across the cylinders, a too-lean fuel mixture, or a prolonged climb at improper power settings or airspeed. There are three steps you should take if the temperature gauge climbs out of limits:

1. Enrich the fuel mixture.
2. Fly level at 55% power.
3. Open the cowl flaps.

■ In-Flight Engine Restart

The two most common reasons for inadvertently killing an engine in flight are running a tank dry and pulling the mixture control when all you really wanted to do was apply carburetor heat. To protect the engine from a sudden burst of power as you refire the motor, follow a logical restart procedure:

1. Throttle—retard.
2. Fuel mixture—full rich.
3. Fuel selector—switch tanks.
4. Fuel pump—on.
5. Restart engine—normal procedure.
6. After restart—fuel pump off, reposition throttle and mixture.

■ In-Flight Engine Fire

The best defense against the very rare occurrence of an engine fire aloft is a thorough preflight inspection. When you lift the cowl, look for pooled or caked engine oil, which indicates a leak. Look, too, for the patch of dried fuel dye that indicates gasoline is dripping where it shouldn't. Inspect the exhaust manifold for the powdery white lines that indicate a hairline crack.

Should you ever experience an engine fire, remember that you want to keep additional fuel from supplying the fire, reduce the smoke entering the cockpit, and descend rapidly to the nearest reasonable landing area.

If your flight examiner calls for a demonstration, follow a checklist:

1. Cabin heat vent—close.
2. Fuel selector—off (simulate only, on the flight test).
3. Fuel mixture—lean to cut-off (simulate only on the flight test).
4. Throttle—close.
5. Magnetos and master switch—off (simulate only, on the flight test).
6. Landing gear—manually extend (simulate only, on the flight test).
7. Emergency descent—cruise speed or maximum gear extension speed.

■ Manual Landing-Gear Extension

Flight examiners do not ordinarily ask to see an in-flight demonstration of manually extending the landing gear. If your plane is a retractable, however, expect questioning on the procedure. Follow a checklist:

1. Landing-gear motor — pull circuit breaker or remove fuse.
2. Altitude — 2000 feet AGL minimum if conditions permit.
3. Landing-gear switch — down.
4. Airspeed — reduce to gear extension speed.
5. Manually extend gear — aircraft flight manual procedure.
6. After landing — prevent ground retraction: leave gear-motor circuit inoperative, leave gear switch down.

While manually extending a landing gear is an approved procedure, manually *retracting* a gear is not. If, after takeoff, the landing gear does not come up with the gear switch, something is obviously wrong. Leave the gear extended until you can land and get the thing fixed. (Make sure you return the switch to "down.")

Also, be aware that some planes place the manual extension handle in an inconvenient position. Do not let this inconvenience deter you from looking for traffic as you work with the crank. Look for traffic as you crank the gear. A mid-air could render your gear extension labors totally useless.

■ Runaway or Inoperative Electric Trim

If an electric trim should either run beyond your control or does not move at all, use a checklist:

1. Aircraft attitude — physically hold enough control pressure to override adverse trim to maintain proper attitude.
2. Deactivate trim system — pull circuit breaker, remove fuse, trim switch off.
3. Re-trim — use manual trim control.
4. Electric trim circuit — leave inoperative until repaired.

■ Electrical System Overload

Some ammeters display the electrical load on the alternator. Others show whether the battery is charging or discharging. Prior to your flight test, review your flight manual to determine the normal indication for your plane.

Many light airplanes are also equipped with a circuit-overload malfunction warning light. Should the warning light illuminate in flight, switch off the entire master switch for a few seconds to allow the alternator to reset. If the overload condition does not reoccur when the master switch is turned back on, continue the flight. If, on the other hand, the warning light returns, an alternator malfunction is likely. In this case, turn off the alternator half of the master switch and head for an appropriate airport for repairs, since the airplane's electrical needs may drain the battery. Don't worry about the engine quitting even if you do drain the battery — aircraft engines receive their firing power from the magnetos, not the battery.

■ Failing Oil Pressure

If you note a substantial reduction of oil pressure in flight, look at the oil *temperature* gauge to analyze the situation. If oil temperature remains normal, the malfunction most probably lies in the oil pressure gauge itself. In this case, maintain normal engine power and divert to a nearby airport to evaluate the source of trouble.

If, however, the oil pressure gauge shows near-zero pressure *accompanied by* high oil temperature, you must anticipate an imminent engine failure. Reduce power to below 55% and use the power that remains to reach a nearby, suitable landing area.

EMERGENCY LANDINGS

Fortunately, engine failure in flight that would require an emergency landing is rare. If, for example, you asked 200 pilots how many have had a forced landing, I doubt that there would be more than one or two.

Even though an emergency landing will probably never be necessary, you must fly prepared for the eventuality. Even *one* botched forced landing is enough to spoil your whole flying career. And remember, none of us ever become expert at making an emergency landing. Therefore, your plan of action must be easy to remember and it must be a plan with *wide* margins for error.

During your flight test, the examiner may reach over to close the throttle and say, "Now, let's see your emergency landing procedure." Your response should follow a simple five-part checklist:

1. Establish best glide speed.
2. Select a landing area.

3. Fly directly to the landing area.
4. Try a restart, communicate, prepare the occupants.
5. Fly a square pattern and land.

■ Establishment of Best Glide Speed

The aircraft's flight manual states (under "emergency procedures") the glide speed that delivers the greatest distance. If you don't know this speed when you need it, don't take time to look it up—just apply a rule of thumb: for fixed-gear aircraft—your glide speed should be one-third greater than the lowest speed in the airspeed indicator's green arc; for retractable-gear aircraft—your glide speed should be 50% greater than the lowest speed in the airspeed indicator's green arc.

Establishing best glide speed as the first step is paramount. Both altitude and distance are precious commodities; therefore, it is important that you *trim* to best glide speed, and then, should anxiety distract you from accurate flying, the plane itself will be delivering its best for you.

■ Selection of Landing Area

Most light aircraft will glide about 2 miles for each 1000 feet AGL; this range assumes no wind and an accurately flown airplane. However, there may be winds and distractions may affect your piloting skill. Therefore, pick the best-looking area that lies within 1 mile for each 1000 feet AGL you are flying. If, however, you are within 2000 feet AGL when the situation arises, modify the procedure by confining your selection of a landing area to one that requires no more than a 30-degree change of direction to get there. Turns requiring a larger change in direction require more altitude than you might realize.

Once you have selected your landing area, don't lose sight of it. Inadvertent turns while you are looking elsewhere may cause you to lose track of its location.

■ Direct Flight to Landing Area

Do not attempt to lose excess altitude with S-turns as you fly toward your landing area—doing so requires experience and a keen judgement that may not be yours during the excitement of the moment. If there is excess altitude, lose it *after* you are over the landing area, with a series of descending 90-degree turns.

■ **Restart Attempt, Communication, Preparation of the Occupants**

If time and conditions allow, try to restart the engine:

1. Carburetor heat — hot.
2. Fuel selector — switch tanks.
3. Mixture control — full rich.
4. Throttle — advance.
5. Electric fuel pump — on.
6. Magnetos — check each.

Try to transmit a distress call if time and altitude permit. If you are already working a frequency, just transmit to those folks; otherwise, use 121.5. Your distress call should include 4 items:

1. "Mayday — Mayday — Mayday."
2. Make and "N" number of your plane.
3. Advise that a forced landing is underway.
4. Approximate the location of the landing area you have chosen.

You would like to know that help is on the way.

Let your passengers know the situation. People tend to remain calmer if they know what's going on and what you are doing about it. Point out your landing area, reassure them that you will reach it with a workable landing, and let them know that help will be on the way once you are down. Put your passengers to work. Assign one the task of unlatching the door to keep it from jamming shut with a hard landing. Give another the task of stowing any hard objects on the floor. Another can be assigned the responsibility of confirming that all seat belts and harnesses are snug.

■ **Square-Pattern Flight and Landing**

Descend around the perimeter of the area with a series of 90-degree turns. These turns should place you on a slightly-higher-than-normal half-mile final. On turning final, quickly determine whether or not an extension of your glide path will reach the area's mid-point. (Evaluate this by the *apparent motion* of a mid-point landmark.) If you have the mid-point made, you know that you can use flaps or a slip to land on the near end of the area.

Once you have committed to the landing, run a short, forced-landing checklist:

1. Flaps — extend as needed, or slip.
2. Landing gear — down and locked.
3. Fuel mixture — lean cut-off.
4. Fuel selector — off.
5. Magnetos — off.
6. Master switch — off.

You should touch down in a nose-high, slow-speed, soft-field attitude. Use positive braking to minimize the landing roll and use nose-wheel steering to avoid hitting anything solid. (Maintaining directional control may require a real effort. Once the nosewheel is down, the rough surface will try to bounce your feet from the rudder pedals.)

Once the plane is stopped, quickly get your passengers out and away from the plane until you are sure there is no prospect of fire.

If I were asked to state only three golden rules for handling a malfunction aloft, I'd say:

1. Communicate with the ground at your first feeling of real concern, *before* the situation becomes an emergency.

2. Know your plane's flight manual as well as you know the lyrics of your favorite song. Keep the manual aboard. Modern flight manuals suggest remedies for most common malfunctions.

3. Level with your passengers. Tell them the nature of the problem and what you intend to do about it.

Follow these simple golden rules and you should have minimal trouble coping with the problem.

FLIGHT-TEST GUIDELINES

■ Equipment Malfunctions

Your response to the flight-examiner's simulated emergency situation should follow a logical checklist.

Rough-Running Engine

1. Clear for traffic.
2. Fuel mixture — full rich.
3. Electric fuel pump — on.

4. Fuel selector—switch tanks.
5. Throttle—restore cruise power.
6. Carburetor heat—full hot.
7. Magnetos—check each, return to "Both."
8. Use smoothest throttle, mixture, ignition.
9. Communicate with a ground facility if conditions persist.

Overheating Engine

1. Fuel mixture—enrich.
2. Power—reduce to 55%.
3. Cowl flaps—open.

In-Flight Restart

1. Clear for traffic.
2. Throttle—retard.
3. Fuel mixture—rich.
4. Fuel selector—switch tanks.
5. Fuel pump—on.
6. Restart engine.
7. After restart—fuel pump off, reposition throttle and mixture.

In-Flight Engine Fire

1. Clear for traffic.
2. Cabin heat vent—close.
3. Fuel selector—off.
4. Fuel mixture—lean to cut-off.
5. Throttle—close.
6. Magnetos and master switch—off.
7. Landing gear—manually extend.
8. Emergency descent—cruise speed or maximum gear extension speed.

Manual Landing-Gear Extension

1. Clear for traffic.
2. Altitude—2000 feet AGL minimum.
3. Landing-gear motor—pull circuit breaker or remove fuse.
4. Landing-gear switch—down.
5. Airspeed—gear extension speed.
6. Manually extend gear—per aircraft flight-manual procedure.
7. After landing—leave gear extension system inoperative until repaired.

Runaway or Inoperative Electric Trim

1. Aircraft attitude — physically override adverse control pressures.
2. Clear for traffic.
3. Deactivate electric trim system.
4. Re-trim — use manual trim control.
5. After landing — leave electric trim system inoperative until repaired.

Electrical System Overload

1. Master/alternator switch — cycle to reset.
2. Alternator switch — leave off if conditions persist.
3. Electrical accessories — reduce electrical load.

Failing Oil Pressure

1. Oil temperature gauge normal — continue flight to nearby airport.
2. Oil temperature gauge overheating — seek immediate safe landing area.

■ Emergency Landings

1. Establish best glide speed.
2. Clear for traffic.
3. Select a landing area — within 1 mile per 1000 feet AGL (within a 30-degree heading change if below 2000 feet AGL).
4. Fly directly to the landing area — at best glide speed.
5. Try restart, communicate, prepare the occupants.
6. Excess altitude — lose over the landing area.
7. Fly square pattern and land — magnetos, master switch, fuel flow — off.

6.

Turn Maneuvers

Any student pilot attempting turns for the first time must overcome two difficulties. First, there is the immediate problem of maintaining altitude as you bank the wings into the turn. The airplane tends to lose altitude during a turn because of the *vertical component of lift* and the *load factors*. Second, the ailerons and rudder do not automatically coordinate their displacements. The pilot must balance the control pressures to keep the ball centered throughout the turn. Both of these difficulties can easily be overcome with a little forewarning and information.

So, let's take a few minutes before the flight test starts to look at each source of difficulty in turn. Then the two problems will be half laid to rest before you even take the controls.

VERTICAL COMPONENT OF LIFT

In straight and level flight, the force of lift (acting perpendicular to the wing span) directly opposes the force of gravity (acting directly downward to Earth). Thus, a wing that produces 2000 pounds of lift maintains the altitude of a 2000-pound airplane (plus a little more for the tail's aerodynamic down-load). When turning, however, lift no longer *directly* opposes gravity. When the wings bank, the direct force of lift is deflected from its alignment with the force of gravity. You then have only the diminished *vertical component of lift* supporting the airplane (Fig. 6-1). Naturally, the plane wants to lose altitude.

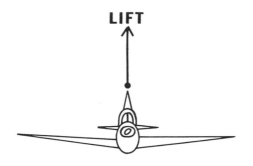

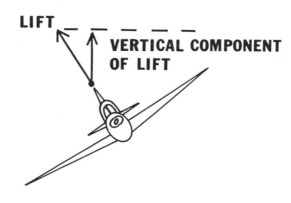

Fig. 6-1. Banking the wings diminishes the vertical force of lift.

LOAD FACTORS

The circular path of a turning plane creates centrifugal force, which adds weight to your airplane. This additional weight is called a load factor, with normal weight represented as 1.0 G (1.0 gravity). These load factors are determined by the angle of bank, and are predictable (see Fig. 1-5)—the greater the bank, the greater the load factor. As the figure shows, the load factor is negligible in very shallow turns, but it accelerates dramatically as you pass through 40 degrees of bank.

There are two forces at work, each trying to make the turning plane lose altitude—diminished vertical lift and increased weight. The solution, of course, is to create additional lift in the wings. And, the most sensible way to do this is to increase the *angle of attack* of the wing. Therefore, when you turn you can expect to pitch the nose up a bit.

The centrifugal force is also responsible for your need to balance the movements between rudder and ailerons.

COORDINATING RUDDER AND AILERONS

To stay on the track, a race car depends on turns that are banked to match the track's radius of turn and the car's rate of turn. Imagine a 200-mph Indianapolis race car going into an unbanked turn—centrifugal force causes the car to skid to the outside, and off the track. The same thing happens to a plane when the bank is insufficient for the turn—you go off the "track" of your flight path.

Or, imagine a 40-mph Volkswagen trying to negotiate the high-banked turn of a raceway. Here, the bank is too great for the car's centrifugal force, and it slips to the inside. The plane behaves the same way, with the angle of bank creating the centrifugal force.

It is necessary then, when turning, to coordinate ailerons and rudder in such a manner that you match the bank and rate of turn with the centrifugal forces at hand. One instrument, the *ball,* gives a quick indication of this coordination (Fig. 6-2).

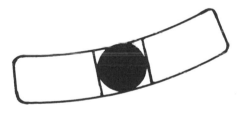

Fig. 6-2. A centered ball indicates good coordination between ailerons and rudder.

■ Coordinated Turn

In a coordinated turn the angle of bank produced by ailerons exactly matches the rate of turn produced by the rudder. Two *favorable* situations exist when the angle of bank and the rate of turn are in balance.

First, the force of lift is in alignment with the line of centrifugal force (Fig. 6-3). Notice how the ball is also in alignment. This happens

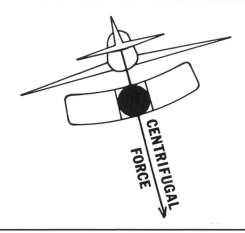

Fig. 6-3. With lift and centrifugal force in alignment, the ball is centered.

simply because centrifugal force positions that ball in its liquid-filled race.

Second, the airplane's fuselage is tangent to the turn, with the nose pointing neither inward toward the turn, nor outward away from the turn (Fig. 6-4). The plane is flying most efficiently, producing the least drag, and the slipstream is flowing directly across the wing's curved upper surface.

■ Skidding Turn

When there is insufficient bank for the centrifugal forces generated in the turn, the plane "skids" to the outside. Here, two *unfavorable*

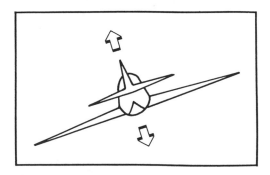

Fig. 6-4. With the fuselage tangent to the turn, the plane delivers its best efficiency.

situations exist. First, the force of lift is no longer in alignment with the centrifugal force. Figure 6-5 shows that the centrifugal force also pulls the ball out of alignment. The ball, deflected to the outside of the turn, tells the pilot that a skid is occurring (and the slipstream no longer flows directly across the designed wing curvature).

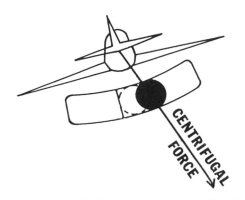

Fig. 6-5. The centrifugal force of a skidding turn pulls the ball away from the direction of turn.

Second, when the plane skids, the fuselage is no longer tangent to the turn—the nose is offset slightly into the turn, and the tail is offset away from the turn (Fig. 6-6). This sidewise flight through the air, as the side of the fuselage mushes through the air, develops unwarranted drag. The plane slows appreciably and loses altitude.

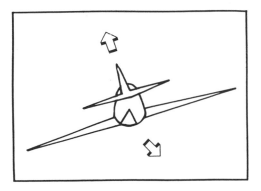

Fig. 6-6. The fuselage is no longer tangent to the turn during a skidding turn.

A skid means you entered the turn using too much rudder. To correct the problem, either apply some rudder pressure in the direction of the ball's displacement, or increase the bank. For example, in a left-skidding turn (Fig. 6-6), you would use some right rudder, or increase the bank. This would correct the initial rudder error, as well as swing the plane tangent to the turn's circle.

■ Slipping Turn

A turn with an angle of bank too great for the centrifugal force generated in the turn is called a "slip"—the plane slips to the inside of the turn. Just as in the case of the skid, two *unfavorable* situations exist. Again, the force of lift is no longer in alignment with the centrifugal force. Figure 6-7 shows that inadequate centrifugal force allows the ball to fall out of alignment. The ball, deflected to the inside of the turn, tells the pilot that a slip is occurring.

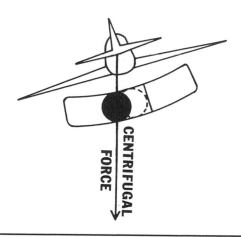

Fig. 6-7. The slipping turn with inadequate centrifugal force allows the ball to fall toward the direction of turn.

Just as with the skid, the fuselage is no longer tangent to the radius of the turn. This time, however, the plane's nose is offset away from the turn with the tail deflected into the turn (Fig. 6-8). This sidewise flight also adds drag, reduces speed, and loses altitude.

A slip means the turn was entered with insufficient rudder for the angle of bank. To correct the problem, just apply rudder pressure in the direction of the ball's displacement, or decrease bank. (An old flying axiom says, "step on the ball" as a quick reminder of needed rudder

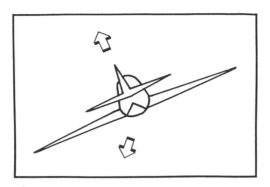

Fig. 6-8. The slipping turn allows the plane to fly sidewise, with the tail drifting to the inside of the turn.

pressure. Just apply rudder in the ball's direction of displacement, regardless of whether contending with a skid or a slip.) In the left-slipping turn (Fig. 6-9), use some left rudder to correct the situation, which corrects the initial rudder error, as well as swings the plane tangent to the turn's circle.

MAKING THE TURN

Think of a turn as a three-stage maneuver: (1) entering into the turn, (2) maintaining the turn, and (3) recovering from the turn.

Entering into the Turn

Before entering into a turn, the simple, emphatic, number one rule is to *clear the sky for other aircraft in the direction of turn*. The examiner's tone may turn a bit agitated if you do not look before you turn. Let me say this agitation is justified. To turn without first clearing, places everyone involved in a terminally hazardous situation.

Rudder and aileron pressures are needed only to enter the turn and establish the desired angle of bank. Once the desired bank is attained, neutralize both rudder and ailerons. The plane will continue to turn until you take positive opposite recovery action with the controls.

Establish your desired bank by using two visual references. First, learn by trial and error, how to "eyeball" the angle of bank against the horizon. Remember though, you must sit straight in the seat to visually gauge the correct angle. Resist the tendency to lean away from the bank when turning. Trying to gauge the angle of bank from a skewed viewpoint is next to impossible.

Second, confirm your estimate with the angle of bank index on the *attitude indicator* (Fig. 6-9).

Fig. 6-9. Confirm your estimated angle of bank by referring to the attitude indicator.

As you initially enter into your turn with ailerons and rudder, simultaneously apply back pressure on the yoke to lift the nose. Doing so gives the additional angle of attack needed to replace the diminished vertical component of lift and offsets the increased load factor.

How much do you have to lift the nose? Well, that depends on the steepness of the bank. A little trial and error soon has you doing it, though — and a good visual reference helps. With your fingertip, put a "smudge" on the windscreen right on the horizon line in level flight. This smudge now gives good pitch information during the turn; just hold it as a bull's-eye above the horizon — the steeper the bank, the higher the bull's-eye.

The smudge, directly in front of your eyes, makes a far better point of reference than the center of the nose cowl. Chances are, you sit in the left-hand seat of the cockpit. From this position, the cowl's centerpoint looks different in left and right turns. A left turn has the centerpoint looking too high, while a right turn has it looking too low (Fig. 6-10). However, your smudge, directly in line with your eyes, gives the true picture of pitch.

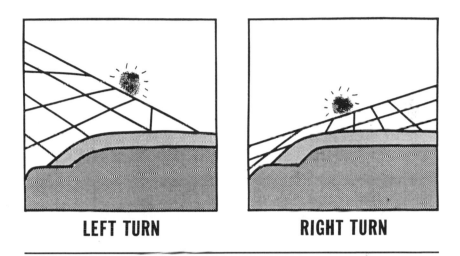

LEFT TURN   RIGHT TURN

Fig. 6-10. A smudge, placed on the windscreen at the horizon line in level flight, helps gauge the required nose pitch when turning.

■ Maintaining the Turn

Although aileron and rudder pressures are not *held* during the turn, they are often *momentarily* needed to maintain the desired bank in the turn. Once you have established your desired angle of bank, the natural ripple action of the air, the inherent stability of the aircraft, or allowing the pitch attitude to vary will disturb the slant of your wings and the angle of bank.

Ripple Action of Air

Air is not perfectly smooth. The uneven ripples within the air tend to move the wings. You may need to nudge coordinated aileron and rudder pressure to regain the desired bank.

Aircraft Stability

The inherent stability of the aircraft attempts to level the wings in a shallow-banked turn. (A 15-degree bank is considered shallow in a light training aircraft.) You will find that a prolonged shallow turn requires frequent renewal of aileron and rudder pressure to re-establish the desired bank.

A medium-banked turn (about 30 degrees) on the other hand seems to have a stability of its own. The plane in this bank tends to hold the bank, and very little corrective control pressures are needed.

A steep-banked turn (45 degrees), however, brings instability to the turn. The plane tries to steepen the bank further, and you will find that opposite, momentary rudder and aileron pressures are needed to keep the bank where you want it.

Pitch Attitude

If you allow the nose pitch to vary in the turn, the bank will also vary. Increase pitch and the bank steepens; decrease pitch and the bank shallows. If you see that your pitch has changed—with a resulting error in bank—just return the nose to its proper position and the bank corrects itself.

Similarly, allowing pitch to move up or down causes the plane to either gain or lose altitude in the turn. If the nose gets too high in the turn, the plane climbs; if the nose dips, the plane loses altitude—just as in straight and level flight.

■ Recovering from the Turn

Whether turning to a landmark or point on the heading indicator, lead your roll-out by half the degrees of bank that you are holding in the turn. For example, if your bank is 30 degrees, then begin leveling your wings 15 degrees before reaching your target. This, naturally, prevents rolling *past* your intended roll-out heading.

Initiate your roll-out by applying opposite rudder and ailerons to level the wings, while you *simultaneously* lower the nose pitch to the level flight attitude. If you delay lowering the nose until the wings level, you will gain 50 or 60 feet of unwanted altitude.

Once the wings and nose regain straight and level flight, neutralize the rudder and ailerons, and execute your final responsibility of the turn—immediately clear for traffic.

Let's quickly review the eight simple steps:

Step 1. Clear for traffic.

Step 2. Lift the nose while applying aileron and rudder pressure to establish the desired angle of bank.

Step 3. Once the desired bank is established, neutralize aileron and rudder pressure, holding back pressure on the yoke to maintain the slight nose-high pitch attitude.

Step 4. Throughout the turn, apply any slight, coordinated aileron and rudder pressure required to maintain the desired bank. Maintain back pressure on the yoke.

Step 5. Begin the turn recovery half the degrees of the turn prior to reaching the desired recovery heading.

Step 6. Accomplish the roll-out with coordinated aileron and rudder pressure, while lowering the nose to the level flight attitude.

Step 7. Once level, neutralize the rudder and ailerons.

Step 8. Clear for traffic.

COMPASS ROSE AND AIRCRAFT HEADING

The ability to roll out on a specific heading implies that student pilots know and understand the compass rose (Fig. 6-11). Unfortunately,

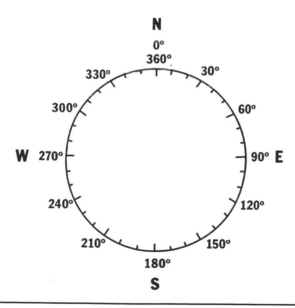

Fig. 6-11. The compass rose shows the degree readings for each compass direction.

often neither implication is true. Most students have never used a compass until they climb into the cockpit, and the numbers on it are meaningless figures.

However, these students have three things going for them when it comes time to learn the compass rose. First, they know the cardinal directions (North, South, East, West) by name and sight, and they have been using the directional lines on the ground for years. Most streets, for example, run in these directions. Second, over most of the country, the majority of roads, property lines, even rows of crops, are laid out in these cardinal directions and are visible from the air. The patterns are quite apparent (Fig. 6-12). Third, almost any student can look at the ground patterns below and eyeball any given angle (say, 30 degrees) from a cardinal line on the ground.

Compass directions should be thought of merely as angles from a cardinal line. The compass rose has 360 degrees, 1 degree for each degree in a circle. The rose starts at North (either 0 or 360 degrees) and continues clockwise. Thus, a right angle from North is at 90 degrees, East, with the northeast compass points lying between 0 and 90 degrees. A

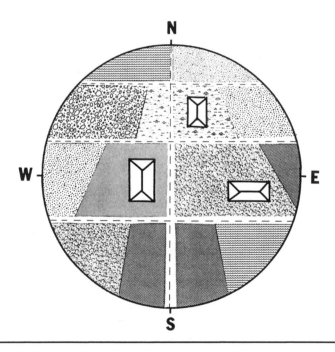

Fig. 6-12. Many sections of our country are "quilted" with a pattern of North-South, East-West lines.

second 90-degree angle from North is at 180 degrees, South, with the southeast points between 90 and 180 degrees. The third right angle is at 270 degrees, West, and the last 90-degree angle is 360 or the starting 0, back North. The beginning and end of the compass rose (0 or 360 degrees) is the same point.

Any eyeballed line from any one of the cardinal lines can become a compass direction. A heading of 60 degrees, for example, becomes an angle pictured as 30 degrees to the left of any nearby East line; a 260 heading becomes an imagined line running 10 degrees to the left of a West line.

It is important to learn how to *picture* aircraft headings — the pilot who does so can easily relate to the terrain, picture how the airplane is traveling across it, and rarely gets lost. If you just *mechanically* turn to a numerical value on the heading indicator, however, you will constantly wonder exactly where you are. Learn to picture directions. (You can even learn this skill while in your car by carrying along the illustration of the compass rose. When you are driving a street of known direction, turn the compass rose around to align it with your direction of travel. Then, call a compass point to yourself, and point to it. It works.)

HEADING INDICATOR

Let the heading indicator reinforce your picture of the roll-out heading, and increase your accuracy. Notice how the instrument *shows* your position relative to the ground directions below (Fig. 6-13). The illustration has us flying a southeast heading of 120 degrees. If we want to turn to 330 degrees, the instrument tells us our destination lies 30 degrees to the left of the North lines, and behind our left wingtip.

When turning to a specific heading, use a combination of an "eyeballed line" and the heading indicator to follow these steps:

1. Use the *heading indicator* to confirm which cardinal line on the ground lies closest to your desired roll-out heading.
2. Using the face of the instrument to assist with accuracy, "eyeball" an angle from that line and select an appropriate landmark to estimate your roll-out heading.
3. Begin turning toward the landmark.
4. As the turn closes with the landmark, compare your estimated "eyeball" value with the heading indicator value. Adjust your estimate so that the plane rolls out on the heading indicator value of the roll-out heading.

Fig. 6-13. Let the heading indicator assist your visualization of the roll-out heading.

To further your skill with the compass rose and turns to headings, apply two rules of thumb:

1. When leaving an existing heading, never just arbitrarily turn left or right. Turn left or right to a *specific* predetermined heading.
2. Never leave an existing heading unless you know to what specific heading you wish to turn.

Come to think of it, those are not bad rules to follow when considering a change of direction in your life—a lot in flying works pretty well in living.

IN REVIEW

■ A combination of *load factor* and reduced *vertical component of lift* can lead to altitude loss in the turn.
■ Load factors increase with a bank increase.
■ Clear for traffic before turning.
■ When turning, an increase in angle of attack (nose-up pitch) is needed to offset the increased load factor.

Ailerons and rudder are used to balance the centrifugal force and rate of turn.

Correct an uncoordinated turn by applying rudder in the direction of the ball's deflection.

Once the desired bank is established in the turn, neutralize ailerons and rudder.

Maintain the needed pitch-up attitude throughout the turn.

Maintain the desired angle of bank throughout the turn with coordinated ailerons and rudder.

Lead the turn's roll-out by half the number of degrees in the bank.

Lower the pitch attitude as you roll out of the turn.

Clear for traffic immediately after completing the turn.

Learn to *picture* aircraft headings as they relate to the terrain.

FLIGHT-TEST GUIDELINES

■ Turn Maneuvers

Perform your flight-test demonstration as a series of eight steps:

1. Establish a working altitude no lower than 1500 feet AGL.
2. Execute a 90-degree clearing turn.
3. Establish an entry heading and confirm recovery heading by referring to heading indicator and landmarks (the examiner will specify either 180 degrees or 360 degrees of turn).
4. Roll into a 45-degree banked turn with coordinated controls, applying pitch to maintain altitude.
5. Remain geographically oriented, by dividing your attention between the plane and the terrain.
6. Begin the recovery with an adequate lead to level the wings on the specified recovery heading.
7. Use a ground reference and heading indicator to maintain the recovery heading.
8. After roll-out, clear for traffic

Flight-Test Tolerances

Altitude in turn and recovery: $+/-$ 200 feet.
Controls—coordinated throughout.
Recovery heading: $+/-$ 20 degrees

■ Descending Turn

This maneuver demonstrates your ability to descend around a landing position after a simulated engine failure. Perform your flight-test demonstration of a gliding turn as a series of nine steps:

1. Establish an entry altitude at least 2500 feet AGL.
2. Execute a 90-degree clearing turn to confirm absence of conflicting traffic.
3. Locate the best emergency landing site within safe gliding distance (within 1 mile per 1000 feet AGL). Advise your examiner of the site you selected. Confirm the wind direction.
4. Apply carburetor heat and reduce power. Clear the engine with power each 1000 feet.
5. Establish the aircraft's best-glide speed and trim to that speed.
6. Fly directly over the selected site, maintaining best-glide speed.
7. Descend in a coordinated gliding turn around the landing site. Maintain a fairly constant radius, employing the skills of *turns around a point*.
8. With the ability to land assured, roll out of the turn on a landing heading.
9. Recover from your maneuver at a safe altitude of at least 1000 feet AGL. Clear for traffic and initiate a recovery, climb at best-rate-of-climb speed.

Flight-Test Tolerances

1. Airspeed — within 10 knots of the aircraft's best-glide speed.
2. Controls — coordinated throughout.
3. Geographic orientation — maintain throughout.
4. Angle of bank — not to exceed 40 degrees at steepest point.
5. Safe recovery altitude — at a position from which a landing could be made.

7.

Flight by Instrument Reference

The *rudimentary* skills for instrument flying that you are taught as a student pilot in no way prepare you to deliberately take on IFR (instrument flying regulations) weather. Most pilots who have depended on this rudimentary ability to do so have flown themselves and their passengers into grief.

If the premise of the opening statement is correct, why then, are you taught elementary aircraft attitude control by instruments? Well, there are two valid reasons. The first reason is efficiency. Instruments are *aids* to flying by outside visual aids. A VFR (visual flight rules) pilot depends on adequate outside visual reference. These visual references alone, however, do not provide accurate flying. For example, a clear, sunny day does little to hold your altitude. Only by *integrating* the altimeter information with the outside references, can you maintain truly level flight. And, only by comparing the heading indicator reading to the landmark, can you fly a course right on the mark. The short time spent emphasizing control by instrument reference is to give the student the ability to *integrate* visual and instrument references to perform more efficiently.

The second reason for developing elementary instrument flying capability is to provide a means of emergency escape. Many (but by no means all) brief, unintended IFR encounters can be resolved with the basic ability to turn the plane back to VFR conditions—many, many pilots attest to the truth of this statement. However, the best defense against an inadvertent IFR encounter lies in your efforts to avoid them in the first place.

■ AVOIDING IFR CONDITIONS

Your "weather avoidance" procedure should (1) obtain an FSS (Flight Service Station) weather briefing anytime that your proposed flight will carry you away from the immediate vicinity of the airport and (2) establish your own personal weather minimums that *must be met prior* to commencing a flight.

■ Weather Briefing

Years ago, pilots took off not knowing for sure just what kind of weather they might encounter. Now, a thorough FSS weather briefing is as near as your telephone—and, today's forecasts are better than 80% accurate.

Weather systems can develop and move fast. Call for a briefing, even for an extended flight within the local area.

■ Personal Weather Minimums

To protect yourself from unintended encounters with unacceptable flying weather, develop your own personal weather minimums. The minimums prescribed by the FARs won't work because they are the absolute *legal* minimums and in no way are intended for use as *safe* operating procedures. (The FAR 3-mile visibility minimum for VFR flight, for example, is *not* a safe flight value; it is a legal minimum.)

Please understand that the guidelines for personal weather minimums that are discussed here are just that—*guidelines*. The values that *you* establish for your *own use* may vary, depending on your own experience and skill. Let's discuss personal weather minimums as they apply to reduced visibility, low clouds, and extent of precipitation.

Reduced Visibility

Flying through reduced visibility presents three major hazards aloft:

1. Reduced visibility can hide any bad weather that lies ahead and

around. It is so easy to just fly into true IFR conditions that lie ahead unseen. And, even if you *do* turn out of them, you don't know which way to fly, with poor visibility all around.

2. Reduced visibility can cause a collision with other airplanes, tall obstructions, or high ridges.

3. Reduced visibility, if it's reduced enough, can lead to vertigo and disorientation. Aircraft attitude control can become a problem.

When establishing your own minimum visibility required for take-off, allow a margin of safety for further deterioration once aloft. I'd ask you to consider a 7-mile visibility as a minimum because (1) when visibility drops to 7 miles, you should plot a course to the nearest alternate paved airport and plan a heading and time enroute to the touchdown. Keep updating this heading and time enroute as you progress through the reduced visibility. (I specify a *paved* airport, simply because sod runways are hard to find when you are in a tense or desperate situation.) (2) If visibilities should drop to 5 miles, you should turn toward your alternate airport. (3) By the time visibility falls to 3 miles, you want to be at the airport and in a position to land.

Low Clouds

I have a pilot-friend with several thousand hours of pilot time. She says that

> when departing VFR, I want a ceiling of at least 3000 feet. FARs, after all, require that I fly at least 500 feet below the clouds. This has me at 2500 feet at the outset; that's about as low as I want the weather to force me. By the time I'm forced to 2000 feet, I have plans laid to an airport. If it looks like I'm going down to 1500 feet, I head for the alternate airport and I want to be in the pattern before I'm pressed down to 1000 feet.
>
> Then too, I'm aware that other pilots are up there under those same low clouds, and ceilings below 3000 feet squeeze *all* the VFR traffic into a *very narrow* band of airspace. Low clouds often give low visibility . . . any lower, and the chances for a mid-air are just too high.
>
> I do enjoy flying, but the key word here is "enjoy." Crowding the weather is not enjoyable. It is irresponsible, often frightening, and just plain dumb. I like to fly, and I plan to eventually become a very old pilot. I'll pick oldness over boldness every time!

Extent of Precipitation

FSS weather forecasts define the extent of showers and thunderstorms as percentages of the terrain over which precipitation may be expected:

Isolated — very small number
Few — 15% or less coverage
Scattered — 16–45% coverage
Numerous — over 45% coverage

It is very difficult to circumnavigate precipitation when the areal coverage approaches 45%. (Severe thunderstorm activity is considered "high risk" when the coverage exceeds 10% of the area.) A pilot is at risk when flying through reduced visibility with "precip" hidden from view, especially at night. (Additionally, at night even light rain, which allows daytime passage, coats the windscreen to greatly reduce nighttime visibility.)

Establish your *own* personal weather minimums based on your own experience. Values of visibility and clouds, for example, are only numbers written on paper until you experience them for yourself. Ask your instrument-rated instructor to take you aloft on a marginal day and see for yourself what 4 or 5 miles of visibility looks like, what it feels like to fly beneath a low ceiling of only 1500 feet or so. Then, you can judge your own comfort level and base your weather minimums accordingly.

RUDIMENTARY INSTRUMENT CONTROL

Basically, controlling the plane's attitude by instrument reference depends on exercising three pilot instrument-flying skills: (1) instrument scan, (2) instrument interpretation, and (3) pilot action.

Instrument Scan

A pilot should monitor all of the instruments consistently, for scanning the instruments is basic to aircraft control. This is not easy to do, and developing a good scan is the more difficult of the three basic instrument skills. There is a tendency, for example, to lock your attention on the instrument that indicates a correction is needed. If the plane slides off its altitude, for instance, there is a tendency to focus on the altimeter until it is corrected. Resisting this tendency to fixate on an instrument while ignoring the others requires a planned scan.

Straight and level flight infers holding altitude, heading, and cruise power. Plan your instrument scan to monitor the plane's three primary functions in the order of pitch — compare the altimeter to the nose dot on the attitude indicator; roll — compare the heading indicator to the wings on the attitude indicator; and power — confirm that the tachometer displays cruise power.

There is a self-test for scanning skill. If you *suddenly see* a large error on an instrument, your scan was flawed. For example, a 100-foot error on the altimeter did not *suddenly* occur. The error was there at 20 feet off—then 50—and finally you saw it at 100 feet off. Your scan just did not see it developing.

■ Instrument Interpretation

Once you have seen an instrument's reading, you must interpret its meaning; what is it telling? A low reading on the altimeter, for example, tells you that you need to lift the nose. But the pilot's interpretation of the reading could be in error. How? Let's say the pilot is maintaining 2300 feet with the altimeter's long hand resting on the 3. But then the plane slips 50 feet lower with that long hand now pointing upward toward the 2. In an instant of confusion, a pilot might attempt to push the hand back down to the 3 by depressing the plane's nose. Interpreting the bank attitude by referring to the attitude indicator is susceptible to error, because the instrument's *horizon* banks, rather than the wings.

If you see an error and take corrective action, but the error gets *bigger,* then your interpretation was wrong. Your interpretive skills will improve with experience.

■ Pilot Action

Once the pilot has seen an instrument error through scanning and has interpreted the needed correction, action must be taken. Many pilots fail to deliver this third basic element of instrument control in timely fashion. For some, it is simply because they are content to accept small errors, that is, to leave well-enough alone. But small, easily corrected errors tend to grow into larger problems that are difficult to fix. A 50-foot error in altitude, for example, is easily corrected. However, let that error grow to 100 feet, and it takes real work to correct it. Combine that altitude error with a 20-degree heading error and fixing it really becomes complex. When looked at from this viewpoint, accurate flying is easier than sloppy control.

■ RUDIMENTARY INSTRUMENT FLYING

Your flight examiner will ask you to perform several simple maneuvers with only instrument references to guide you. Each maneuver is designed to provide a means of escape from IFR conditions.

As you perform these maneuvers, you must keep one fact firmly in mind: you are not an instrument pilot. Your instrument capability is rudimentary, at best. Therefore, keep the magnitude of your maneuvering small. Keep the attitude of the plane as close to straight and level as practical and just change the pitch, roll, or power enough to accomplish the maneuver in "slow motion."

You want to maintain a near-level attitude for the following reasons:

1. A high-pitch attitude can lead the inexperienced instrument pilot into a stall or a turn that develops into a spiral. Recovery from a developed spiral is highly questionable.

2. A nose that is too low can lead to a high-speed turn that also develops into a spiral.

3. A too-steep bank can allow the nose to lower into a spiral.

4. The torque of excess power can produce a turn. This, without the guidance of visual references, inevitably leads to lowering the nose into a spiral.

So, you want to fly the prescribed instrument maneuvers carefully.

■ Tools of Maneuvering

Any time you maneuver the plane, you do so with three basic tools—pitch, roll, and power—in a manner that allows the prescribed instrument maneuver yet disturbs the natural balance of the aircraft as little as possible.

Pitch

Use the nose dot of the attitude indicator to control your pitch. To prevent excess movement of the pitch axis, confine your pitch deflections to three positions as you fly the instrument maneuvers:

1. Hold the nose level against the artificial horizon (Fig. 7-1).
2. Deflect the nose one-half dot upward (Fig. 7-2).
3. Deflect the nose one-half dot downward (Fig. 7-3).

Roll

Use the attitude indicator to control bank attitude. Again, in an effort to preserve the plane's natural stability, position the wings in three attitudes on the instrument:

1. Wings level (Fig. 7-4).
2. A 10-degree bank (Fig. 7-5).
3. A 15-degree bank (Fig. 7-6).

Fig. 7-1. Attitude indicator showing the nose held level against the artificial horizon.

Fig. 7-2. Nose deflected one-half dot high.

Fig. 7-3. Nose deflected one-half dot low.

Fig. 7-4. Wings level against the attitude indicator's artificial horizon.

Fig. 7-5. A 10-degree bank is a useful tool with which to correct small errors in heading.

Fig. 7-6. A 15-degree bank is useful when a change in heading is desired.

Power

Throttle variations normally used in maneuvering upset the plane's pitch moment. When flying VFR your outside vision allows you to easily correct for the imbalances created. Under the hood, however, compensating for pitching moment is not so easy. Confine your power applications to three throttle settings that will accomplish the instrument maneuver, yet disturb the plane as little as possible:

1. Cruise power (use whatever you've been cruising at for the past several minutes) (Fig. 7-7).
2. Cruise power plus 100 rpm (Fig. 7-8).
3. Cruise power minus 200 rpm (Fig. 7-9).

Fig. 7-7. Cruise power (here 2400 rpm) is the throttle you have been running for the past few miles.

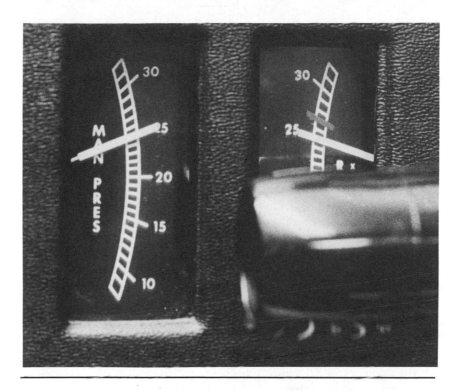

Fig. 7-8. Cruise power plus 100 rpm provides an easy low-performance climb.

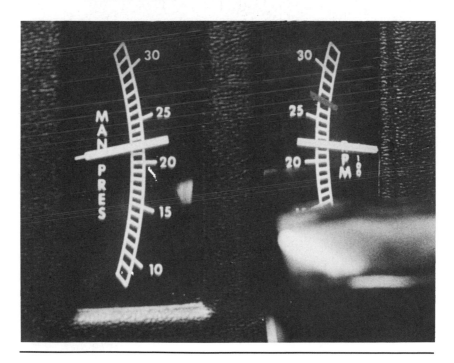

Fig. 7-9. Cruise power less 200 rpm allows an easily managed descent.

FLYING INSTRUMENT MANEUVERS

The flight examiner will ask you to demonstrate five maneuvers "under the hood":

1. Straight and level.
2. Climb.
3. Descent.
4. Turns to headings.
5. Recovery from critical attitudes.

FLIGHT-TEST GUIDELINES

Straight and Level

Straight and level flight consists of a constant series of corrections to altitude, heading, and airspeed. To manage these elements, use the limited-control deflections, discussed earlier, as you correct to maintain straight and level flight.

Altitude

If you see the altimeter slide 100 feet off altitude, apply a one-half nose dot deflection in the appropriate direction. This small pitch change will not disturb your plane, yet will move you back to altitude at about 200 feet per minute (fpm).

Heading

If you see your heading indicator approaching a 15-degree heading error, apply a 10-degree banked coordinated turn in the appropriate direction. This slight bank will not disturb your altitude, yet will return you to course within several seconds.

Airspeed

Leave the throttle at cruise power and accept the small (5 knot) airspeed variations that occur as you apply the one-half nose dot corrections needed to maintain altitude.

Perform the instrument straight and level maneuver as a series of seven steps:

1. Establish cruise speed and power under VFR, while on an easily interpreted 30-degree increment on the heading indicator and on a 1000-

or 500-foot increment on the altimeter.

2. Clear for traffic, then put on the hood.
3. Maintain altitude by holding the nose dot of the attitude indicator astride the artificial horizon.
4. Maintain heading by holding the wings of the attitude indicator level against the artificial horizon.
5. Maintain cruise speed by maintaining cruise power.
6. Confirm with the examiner (now your safety pilot) that the area remains clear of traffic.
7. Maintain straight and level flight.

Flight-Test Tolerances

1. Desired heading: $+/-$ 15 degrees.
2. Desired altitude: $+/-$ 100 feet.
3. Cruise speed: $+/-$ 10 knots.
4. Duration of instrument control: 3 minutes.
5. Control application: coordinated throughout.

■ **Climb**

Perform your "hooded" climb as a series of eight steps:

1. With VFR visual references intact, establish cruise flight at the 1000- or 500-foot increment on the altimeter and on a 30-degree increment on the heading indicator.
2. Clear for traffic, then put on the hood.
3. Increase power by 100 rpm, and raise the attitude indicator's nose one-half dot.
4. Trim to this attitude.
5. Maintain your entry heading. The slight climb power and small pitch attitude used will minimize the left-turning tendency of torque. Small variations in heading, however, will occur. Correct for these variations with a 10-degree banked coordinated turn.
6. Maintain constant climb speed by keeping the nose dot resting atop the horizon line. If the pitch strays from this position, airspeed will change. To re-establish the constant climb speed, just return the nose dot to its proper position and the airspeed will quickly correct itself.
7. Confirm with the examiner that the area remains clear of traffic.
8. Begin your level-off to the assigned height, 50 feet below that altitude:

   a. *Slowly* return the nose dot to its level position.

b. *Slowly* retard the throttle to normal cruise power. Slow readjustment of pitch and power allows the plane to gain those last few feet and accelerate to cruise speed.

c. Complete your maneuver by re-trimming to level cruise flight.

Flight-Test Tolerances

Constant climb speed: 10 knots.
Desired heading: 15 degrees.
Level-off: 200 feet of assigned altitude.
Control usage: coordinated throughout.

■ Descent

Perform your instrument descent as a series of nine steps:

1. Climb to an altitude that will allow a recovery at least 1500 feet AGL.

2. While still flying VFR, establish an entry heading and altitude in cruise flight.

3. Clear for traffic and put on the hood.

4. Begin your descent by first reducing throttle 200 rpm below cruise power. As you do so, maintain a constant pitch attitude that holds the attitude indicator nose dot level with the artificial horizon. This slight power reduction, coupled with a level nose, does little to upset the airplane's natural stability, yet produces a 400-fpm descent, while maintaining airspeed.

5. Trim the aircraft to hold attitude at this new power setting.

6. Confirm with the examiner that the area remains clear of traffic.

7. Maintain your descent heading. The level nose and minimal power reduction minimizes this task. However, if you see the heading indicator approaching a 15-degree error, correct with a 10-degree banked coordinated turn.

8. Maintain a constant descent airspeed. Again, the power/pitch combination makes this task easy. If you see, hear, or feel any airspeed variation, your nose pitch has strayed. Simply return the nose dot to its level attitude, and the airspeed will correct itself within a few seconds.

9. Begin your level-off 50 feet above the assigned recovery altitude. *Slowly* restore cruise power while you maintain the same, good, level nose dot attitude. Conclude your maneuver by re-trimming the plane to cruise speed and power configuration.

Flight-Test Tolerances

1. Desired heading: $+/-$ 15 degrees.
2. Constant airspeed: $+/-$ 10 knots.
3. Recovery level-off: $+/-$ 200 feet.
4. Control application: coordinated throughout.

■ Turns to Headings

Demonstrate your hooded turns to headings as a series of 10 steps:

1. Establish an entry heading and altitude in cruise configuration.
2. Confirm your stipulated recovery heading.
3. Clear for traffic, then put on the hood.
4. Enter the turn with a coordinated 15-degree banked turn. Maintain cruise power and a level nose dot pitch attitude.
5. Maintain a fairly consistent altitude. The 15-degree bank does not greatly disturb the wing's vertical component of lift. However, the nose held level (for ease of interpretation) allows a *slight* descent. If the altimeter approaches a 100-foot error, apply a one-half dot nose deflection in the needed direction for correction.
6. Maintain a consistent angle of bank. The flight test calls for "approximately a standard-rate turn." The 15-degree bank will deliver this approximate value. (An exact standard-rate of turn is one that turns the plane a full 360 degrees in 2 minutes. The exact bank that produces this rate of turn is dependent on the plane's airspeed. The faster you fly, the steeper you must bank. A simple formula figures the exact bank required:

$$\frac{\text{Airspeed}}{10} + 5 = \text{bank for standard-rate turn}$$

Thus a plane flying 100 knots requires a 15-degree bank:

$$\frac{100}{10} + 5 = 15$$

(The arbitrary 15 degrees is close enough for most training aircraft.)
7. Maintain a consistent airspeed throughout turn. A speed slightly lower than cruise speed is normal; the plane slows down in a turn. But, if you see airspeed fluctuations after the turn is established,

nose pitch is probably at fault. Return the nose dot to the horizon line and the airspeed will stabilize.

8. Confirm with your examiner that the area remains clear of traffic.
9. Monitor heading indicator for the approaching recovery heading.
10. Begin your recovery 10 degrees prior to the recovery heading. *Slowly* roll the attitude indicator's wings to level as you hold the nose dot centered on the artificial horizon.

Flight-Test Tolerances

Desired altitude: +/− 200 feet.
Constant airspeed through turn: +/− 10 knots.
Bank angle: approximately standard rate.
Recovery heading: +/− 20 degrees.
Controls: coordinated throughout.

■ Recovery from Critical Attitudes

The examiner wants you to demonstrate that you can recover from critical attitudes while flying the airplane under the hood. To do this, the examiner will take the controls momentarily while you are under the hood, and without exceeding a 10-degree pitch attitude or 45 degrees of bank, will place the plane in either the approach to a climbing stall or the entry to a power-on spiral. Your job, once the controls are returned to you, is to recover to straight and level cruise flight.

On taking the controls, you have a twofold task:

1. By referring to the instruments, you must first determine the critical attitude—is it the approach to a stall or the entry to a spiral?
2. Once you have identified the critical attitude, you must effect a recovery to straight and level that does not allow a stall or excessive load factors.

Approach to a Stall

First, identify the maneuver as an approach to a stall. Look at the airspeed indicator. If airspeed is diminishing rapidly, you are confronted with a recovery from a developing stall. Put a four-step recovery procedure into play.

1. Promptly return the attitude indicator's nose dot to the artificial horizon line. This is to immediately reduce the angle of attack and to prevent a full stall from occurring.
2. Increase throttle to climb power. This increases windflow over the

wing roots, which decreases the stalling speed.

3. Level the attitude indicator's wings to the artificial horizon line. This reduces aerodynamic loading as it increases the wing's lifting efficiency.
4. Once the nose and wings are level, restore cruise power. Fly the plane straight and level for several seconds to complete the maneuver.

Entry to a Power-On Spiral

Again, the airspeed indicator identifies the maneuver. If the airspeed is increasing rapidly, you are confronted with a recovery from the entry to a spiral. Plan a four-step recovery:

1. Promptly reduce power. You must do this as a first step to prevent excess speed from developing.

2. Quickly level the attitude indicator's wings to the artificial horizon line. There must be little delay in this step if you are to prevent excess aerodynamic loading.

3. *After* you level the wings, *slowly* return the attitude indicator's nose dot to level. If you level the nose too quickly, or while the wings are banked, you will add to the aerodynamic loading and could induce a high-speed stall.

4. With wings and nose level, restore cruise power. Fly the plane straight and level for several seconds to complete the maneuver.

Flight-Test Tolerances

Promptly recognize the present critical flight attitude.
Recover with coordinated controls.
Avoid excessive aerodynamic load factors, airspeed, or stalls.

In reference to flying instrument maneuvers, I want to repeat these statements:

1. Legal VFR weather requirements demand visibilities of 3–5 miles depending upon the airspace you occupy. Common sense asks for even higher *safe* minimums.
2. Do not depend upon your rudimentary instrument skill to deliberately take on IFR conditions. Doing so is very risky business.
3. Recognize your basic instrument capability for its intended purposes:

   a. To facilitate efficient VFR flight by *integrated* visual/instrument reference.
   b. To provide a *possible* means of escape from momentary, inadvertent IFR encounters.

Before we leave this chapter, I would like to discuss a charming relationship that exists between pitch, power, airspeed, and altitude in light aircraft flying within the cruise range (between 55% and 75% of power).

Pitch

1. With power constant, each one-half dot nose deflection on the attitude indicator results in a 200-fpm climb or descent.
2. With power constant, each one-half dot nose deflection produces a 5-knot variation in airspeed.
3. With throttle untouched, each one-half dot nose deflection increases or decreases the tachometer reading by 100 rpm.

Power

1. A 100-rpm change in power will deflect pitch by one-half nose dot.
2. With pitch held constant, each 100 rpm changes the airspeed by 5 knots.
3. With pitch held constant, each 100 rpm creates a 200-fpm climb or descent.

I hope that some clear, spring morning you will fly aloft and enjoy seeing for yourself these relationships play out. You will undoubtedly find satisfaction in learning more about your craft and gain a new appreciation of flying with truly integrated visual/instrument references. And this knowledge that you acquire will serve you well—later—as you pursue your instrument pilot rating.

How can these pitch/power/airspeed/altitude relationships apply to IFR procedures? Let's say you are on an instrument approach, and you see that you need to increase your rate of descent by 200 fpm, yet maintain your existing approach speed. Understanding the relationships will tell you exactly what you must do with the pitch and power to achieve the desired glide slope.

In short, however, you should fly only when the weather is good. If you stay out of the sky when the weather is bad, you should have no trouble at all.

8.

Between the Maneuvers

The FAA publication, *Private Pilot Practical Test Standards,* details the specified tasks that you are to satisfy, along with specific criteria for gauging satisfactory performance. But what the flight-test guideline does *not* discuss, is the quality of your flying *between* the requested maneuvers. Yet, the examiner evaluates every moment that the plane is under your control, from preflight inspection to securing the aircraft. *Can* careless flying between the maneuvers adversely affect a satisfactory outcome to your flight test? Most certainly. For example, if you show inattentiveness to the engine by trying to cruise with excess power, or make a few climbs with the mixture set to cruise fuel flow, or allow a repeated lack of traffic awareness to produce a few potential mid-airs, you will have a very hostile examiner on board.

It is imperative, then, to realize that your pilot skills are being judged during every moment of the flight test. Let's discuss several important elements of "flight between the maneuvers."

USE OF CHECKLISTS

The 15,000-hour airline captain who flies every day would not even *think* of performing a cockpit procedure without the guidance and discipline of a written checklist for that procedure. Captains know that an adequate checklist stands between themselves and a potentially danger-ous omission. Flawed memorization and habit are inadequate defenses. Think, then, how much more important it is for you, with less experi-ence and infrequent flying—and in a light-plane cockpit devoid of warn-ing lights and horns that might signal a task undone—to use checklists. The flight-test examiner *expects* you to use written checklists.

Today's light-plane operating flight manuals provide checklists for all anticipated flight procedures (Fig. 8-1). But these lists are often diffi-

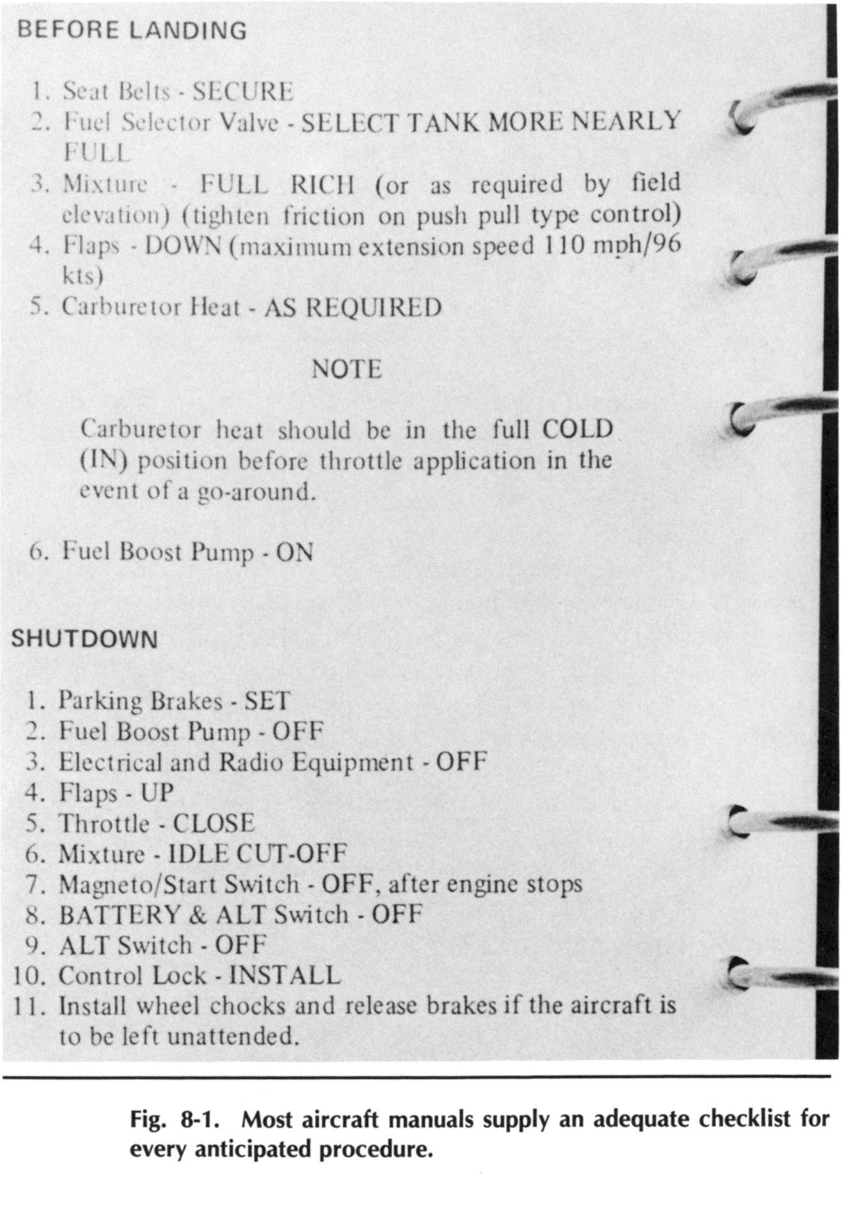

BEFORE LANDING

1. Seat Belts - SECURE
2. Fuel Selector Valve - SELECT TANK MORE NEARLY FULL
3. Mixture - FULL RICH (or as required by field elevation) (tighten friction on push pull type control)
4. Flaps - DOWN (maximum extension speed 110 mph/96 kts)
5. Carburetor Heat - AS REQUIRED

NOTE

Carburetor heat should be in the full COLD (IN) position before throttle application in the event of a go-around.

6. Fuel Boost Pump - ON

SHUTDOWN

1. Parking Brakes - SET
2. Fuel Boost Pump - OFF
3. Electrical and Radio Equipment - OFF
4. Flaps - UP
5. Throttle - CLOSE
6. Mixture - IDLE CUT-OFF
7. Magneto/Start Switch - OFF, after engine stops
8. BATTERY & ALT Switch - OFF
9. ALT Switch - OFF
10. Control Lock - INSTALL
11. Install wheel chocks and release brakes if the aircraft is to be left unattended.

Fig. 8-1. Most aircraft manuals supply an adequate checklist for every anticipated procedure.

cult to find, buried among the many pages. May I offer a suggestion? Reduce the manufacturer's published checklists to your own best penmanship, within the confines of a handy-sized spiral-bound notebook (Fig. 8-2).

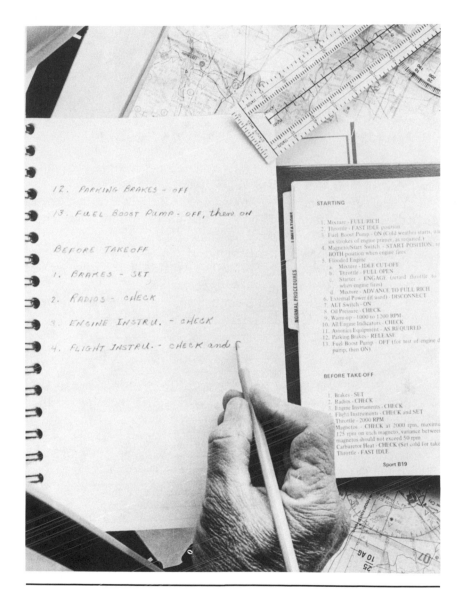

Fig. 8-2. Needed checklists are often difficult to locate within the aircraft manual. Construct your own notebook of checklists.

PRECISION FLYING

Those same applicant-pilots who fly a series of S-turns with the finesse and accuracy of a wheeling seagull often let their between-maneuvers flying wander like a foxhound trying to pick up the trail. Don't let this happen to you. Pay attention to the altitude, heading, airspeed, power setting, and control coordination of your plane *throughout* the examination flight.

■ Straight and Level Flight

Make those moments of cruise flight a showcase of precision flying:

1. Consciously select a 100-foot altimeter increment as your cruise level and maintain that altitude within 50 feet.
2. Quickly establish a heading on one of the 30-degree increments of the heading indicator and maintain that course within 5 degrees, as you wait for your next maneuver assignment.
3. Maintain a power setting within 25 rpm of that recommended for cruise. Airspeeds should vary no more than 5 knots as you make small pitch corrections to maintain altitude.
4. Holding a course often requires small corrections left or right. Even in these miniscule turns, keep the controls coordinated.

■ Turns

Between maneuvers, you may be asked to perform turns to keep you within a designated practice area. Perform these turns as precision maneuvers:

1. Don't just arbitrarily turn left or right to an uncertain heading. Before you begin a turn, determine a definite recovery heading.
2. Make certain that you clear for traffic and use coordinated controls to establish a medium-banked (30-degree) turn.
3. Hold an accurate altitude in the turn; don't let your altimeter stray more than 50 feet from your entry height.
4. Recover from the turn within 5 degrees of your intended roll-out heading.

■ Climbs

There will be several occasions throughout the flight test in which you must climb to a higher altitude to perform a task. Conduct these climbs as precision maneuvers:

1. Initiate the climb along a specific 30-degree increment on the heading indicator—hold that heading within 5 degrees during the climb.
2. Enrich the fuel mixture for the climb.
3. Begin the climb by first elevating the nose to a pitch position that will deliver best-rate-of-climb speed recommended by the flight manual (approximately a full nose dot deflection on the attitude indicator). By increasing pitch prior to adding climb power, you minimize the possibility of over-revving the engine. Trim the plane to climb attitude.
4. After raising the nose, advance the throttle to the recommended climb power (full throttle in many small aircraft).
5. Maintain airspeed as the exact best-rate-of-climb speed by making small (less than one-half nose dot) pitch corrections. Maintain this best-rate airspeed within 5 knots throughout the climb.
6. Once best-rate airspeed is established, re-trim the elevator to hold the correct pitch attitude for the climb power and airspeed.
7. Start to level off 50 feet below your intended recovery altitude. (The plane will gain those 50 feet during the recovery process.)
8. Effect your level-off by *slowly* returning the nose to level. Once the nose is level, *slowly* restore cruise power. This slow process allows the plane to accelerate to cruise speed as the nose levels.
9. With level cruise flight restored, retard the mixture to cruise flow and the trim to level flight. You should recover from the climb well within 50 feet of your desired new altitude.

■ Descent

What goes up must eventually come down. Plan your descent as a precision move.

1. Plan your descent along a definite heading. Again, a clearly delineated 30-degree increment of the heading indicator offers an easily identified course. Stay within 5 degrees of your heading. Decide your desired recovery altitude *before* you start descending.
2. May I offer a pitch/power combination that results in an efficient enroute descent? Use a one-half nose dot downward deflection and 55%

power. This produces a let-down of about 600 feet per minute at a high-cruise airspeed.

When establishing this pitch/power combination, first retard power to 55%, to prevent over-revving as you lower the nose. Know, however, that the tachometer *will* increase as you start to descend. Be prepared to re-adjust the throttle back to 55% power as you continue to descend.

3. Maintain a consistent let-down airspeed with *small* pitch changes. Your airspeed should not vary more than 5 knots during the descent.

4. Begin your level-off 50 feet above your desired recovery altitude. Slowly level the nose as you restore normal cruise power. The plane will lose the remaining 50 feet and accelerate to cruise speed as you do so. Plan a recovery to level flight within 50 feet of the desired recovery altitude.

5. Complete your maneuver by re-trimming to cruise speed and power.

Remain attentive to precision flying between maneuvers — your efforts will not go unnoticed by your examiner.

TRAFFIC AWARENESS

Probably the single-most element of airmanship that many examiners evaluate during your flight test is that of traffic awareness. There are three good reasons for this concern:

1. The examiner is concerned for the safety of *both* of you during the flight test.

2. The examiner is concerned that your early habits will probably set the pattern for your flying career. It's just as easy to set *good* habits and it is safer.

3. The examiner is *mandated* by flight-test procedure to pay special attention to traffic awareness.

You must remain attentive to both actual and potential traffic anytime your plane is moving, either taxiing or aloft. There are, however, several situations that demand special attention to traffic awareness:

■ Prior to Takeoff

If you are departing from an airport without a control tower, you *must* taxi a 360-degree clearing turn before rolling onto the runway. You must be aware of both arrivals and other departures — clear each pattern leg for *every* runway.

■ Departure Climb to Cruise Altitude

If the plane's upturned nose restricts your forward view, make periodic shallow turns to clear the sky straight in front of you.

■ Cruise Flight

You must make a conscious effort to clear every 45 seconds of cruise flight. You can do this by first leaning forward in your seat. This move provides two advantages. First and foremost, the physical move reminds you that you *are* performing a conscious action. Second, the move provides better rearward visibility.

When you look for planes, turn your head toward each 30-degree sector of sky in turn and scan the sector by moving your *eyes,* not your head. Moving your eyes to search each sector will give you an unblurred look of that part of the sky.

Remember, your eyes work like a telescope. Change your range of focus. All too often pilots focus only on the horizon and look right through a plane that is only 3 miles away and closing fast.

■ Changing Altitudes

Any change in altitude calls for a diligent search to clear before adding or reducing power. Look above, below, and to each side. It is so easy to climb right into the belly of an aircraft, or let down right on its cockpit.

■ Changing Headings

You must always look before turning, whether you are taxiing or flying. This is especially important as you turn the legs of an airport traffic pattern, where you have a concentration of aircraft, all headed for the same spot with pilots often engaged in cockpit procedures. To understand this situation, look at the statistics. FAA statistics show that the overwhelming majority of mid-air collisions occur during the approach-to-landing phase of the flight. Please, please, take special care of yourself in the traffic pattern.

■ Radio Communications

You must make a special effort to stay vigilant for traffic during radio communications. Many pilots (unbelievable but true) stare at the radio when communicating. You must make a conscious effort to override this human tendency.

Here is a tip. Don't talk on the radio unless your wings are level. That's right. Don't talk in a turn; wait until you level the wings. This is especially true in the traffic pattern. Traffic pattern turns are close to the ground and you need to pay full attention to manage airspeed, altitude, ground track, and control coordination. The tower can wait a moment for your message, until you level the wings.

■ GEOGRAPHIC AWARENESS

Know your approximate position from your home airport throughout the flight test. This is often not easy to do. You are under the stress of a test, listening to the examiner's demands, and concentrating on performing maneuvers to your best ability while you cover many miles of airspace.

I hope I never forget my experience many years ago when I was giving my first flying lesson as a brand new certified instructor. The student was preparing for the flight test and I was giving the final evaluation ride. The flight soon disclosed that the student still had some problems—all fixable. But the lesson was long and arduous, demanded my full attention, and covered many miles. When the training ended and it was time to go home, I had a problem. I wasn't at all sure which way home was. Did I do or say anything to reveal my foolishness? Heck, no. I was still at the stage of my flying career where I thought I shouldn't admit mistakes. So, I said, "Now we are going to simulate being lost— use your VOR and get us home." And then I sat back and let the student deliver me safely back to the ramp. Flying often requires inventiveness.

Try not to be lost at the conclusion of your flight test. Just a few actions on your part will avoid it.

1. Ask your examiner the initial direction of travel before you even get into the plane. Often the first segment of the flight test involves getting established on the airway toward a preplanned destination. In this case you will already know your departure route.

2. Keep your sectional chart across your lap and keep updating your changes of position.

3. Confirm your changing positions with your VOR during some of your cruise flying between maneuvers.

WEATHER AWARENESS

Always obtain a weather briefing just prior to the flight test. If the local forecast indicates probable weather below your personal weather minimums, discuss this with the examiner. The examiner will not force you to fly in hazardous weather, but will look to you for the go/no-go decision.

Good flying weather can quickly turn bad, even within your local area. If the weather deteriorates once your test is underway, express your concern to the examiner. The two of you will come up with a solution. At worse, you may have to delay the completion of the test to another flight. Or, the questionable weather may still allow the test to be completed within the traffic pattern. Whatever the situation, the examiner wants your input on the weather evaluation. It is an opportunity for you to demonstrate your in-flight, pilot-in-command decision making. You will be making similar decisons on the many flights yet to come, when you fly as a certified private pilot.

Our time together is quickly drawing to a close. If I have discovered anything from my life as a flight instructor it is this: The average student pilot holds the potential for greatness. Please know that I do understand your flying concerns and hope that our time together has helped to sort things out and put these concerns to rest. I feel that you have the potential—that you *will* conduct your flight test to your very best ability. I expect no less from you.

I hope that you will read this book often, jotting down questions or thoughts in the margins—take note of what worked for you in a particular situation. And have some fun with our book if you *do* add copy to it. Give it to a friend someday. There is no finer gift to give than an already-used book, with your *own* thoughts written into the text.

Please do me a favor. Send me a postcard and let me know of your victories or frustrations. I want to hear both. My very best wishes fly aloft with you.

Ron Fowler

Ron Fowler
General Delivery
Christmas, FL 32709
Winter 1993

Suggested Reading

As a pilot, you are duty bound to acquire, study, and understand all significant aeronautical information to which you have access. For as a pilot, you are engaged in one of the few remaining endeavors that allows only *you* to act as the master of your own welfare, as well as the welfare of your entrusted passengers. Once aloft, it is up to you – you are the one that holds the ultimate authority and final responsibility for the integrity of the flight. And it should be no other way.

BOOKS ARE BETTER

It is with this concern for your authority and responsibility in mind, that I gently insist that you build and study your own aviation library. Oh, I realize that there are many aviation videos and audio-cassettes available and that they are *easier* to view and listen to than reading a book. Many provide a useful *supplement* to reading – and they should be just that, a supplement. But I do truly feel that a well-written book is better for transferring knowledge for three reasons. First, a book calls for effort from the reader (and reading is an effort), which helps the reader to better implant ideas, better, certainly, than passively watching or listening to a tape.

Second, a book is better simply because the words of a book can be held in hand for the reader to review, to mull over, to evaluate. Too often, much of the electronic media information is forgotten 20 minutes after the tape is over. Books have staying power.

Finally, a book provides the luxury of leisure; that is, readers can pace the assimilation of knowledge to their own taste. You can stop reading in mid-sentence, take time to think about the idea, put it into the context of your own situation, and then continue. (And the greatest compliment an author can receive is to present an idea, have the reader reshape that idea to their own best needs, then someday return their revision to the writer as a product of their own invention. In other words, a reader and a writer work as a team.) Electronically delivered information, on the other hand, offers little leisure—the viewer or listener must keep pace. If the stream of taped information is interrupted by the listener's own thought, the sequence of information is often lost. The listener falls behind. And from that point to the end of the tape, the viewer or listener must play the frustrating game of catch-up. You can carry a book with you and refer to it at the time you have a question, not wait until you have access to your video or tape player.

I guess what I am talking about is information versus knowledge. Electronics are good at delivering immediate information, but books work better at imparting significant, lasting knowledge. Only the leisure afforded by reading permits you to reflect upon an idea, whenever you want or wherever you are. I feel that reading is your best avenue to knowledge.

AN AVIATION BOOKSHELF

So, I offer a short list of books with which to start your own aviation bookshelf—one that I hope grows as your pilot logbook hours increase.

The Student Pilot's Flight Manual, Seventh Edition, revised
   printing, 1993
William K. Kershner
Iowa State University Press, Ames

Bill Kershner first produced this book over three decades ago. Since then, nearly a million student pilots have learned to fly using the many revised editions. The book, well written and well illustrated, has earned its position as a classic of flight training. Kershner not only tells you how to manage the plane, but also explains the mechanics of preflight planning, communications, weather, and navigation. This book also includes the FAA written test questions (airplanes), answers and explanations, and the practical (flight) test standards. Important reading for any serious flight student, the book has stood the tests of time and repeated use.

Aircraft Operating Manual (For the plane you fly)
Aircraft manufacturer's publication
   There is no such thing as a "forgiving airplane." All planes have

operating limits in terms of airspeeds, aerodynamic forces, loading, and performance. In light aircraft, particularly, the parameters of these limits are quite narrow. Operate the plane within these limits and it will behave and perform as expected—try to operate *beyond* these limits, however, and *no one* knows what to expect. You suddenly become a new test pilot engaged in on-the-job training. The manufacturer's recommended procedures for specific flight situations are stated with the prime intent to keep you within the plane's design limitations.

Not only must you own a copy of your plane's manual, but you must study it to the point that you know its facts as well as you know your own birthday.

Spend a week of evenings studying your plane's flight manual. Then, ask a pilot-friend, with manual in hand, to quiz you on a fact stated on *each page*. Score yourself. You should be satisfied only with at least 9 out of 10 correct answers. If you score anything less, spend another week of study for another quiz.

Federal Aviation Regulations (FARs)
U.S. Government Printing Office publication

I believe I have seen nearly every regulation played out within the cockpit. I can make these statements concerning the FARs:

■ Every FAR makes sense. They should. Before a proposed regulation is put into practice, it is thoroughly debated by both regulators and users. If a regulation later proves unsatisfactory in use, it is quickly modified.

■ While compliance with regulations is no guarantee of safety, every *non*-compliance I have witnessed has put the pilot at risk. Many regulations are not intended to guarantee a *safe* procedure—many have the sole intent to state only a *legal* minimum. (The 3-mile minimum VFR visibility requirement is a good example of this concept. Safe, *perhaps* for close-traffic pattern work, but not too good for cross-country flying.) But I repeat: while compliance offers no guarantee, non-compliance creates risk.

■ Knowledge of and adherence to FARs give order to the sky. Each pilot knows what to expect of the other. Obeying the FARs is comparable to obeying the "rules of the road." Imagine the chaos that would occur on our streets and highways if many drivers did not understand the significance of double lines running down the roadway, or did not know the privilege of the car on the right at a 4-way stop. Chaos would rule the sky without the FARs to set the standards.

Airman's Information Manual
U.S. Government Printing Office publication

Our lives aloft are directed by standard operating procedures. This publication is a concise, explicit guide to these procedures, from pre-flight planning, to communications, to traffic patterns, and beyond. It is well indexed. If there is *any* procedure that confounds you, you will likely find the solution here.

Aviation Weather
U.S. Government Printing Office publication

Who says the government can't write an interesting book? This book is well written and superbly illustrated. Weather is a prime element of nature with which pilots must contend. You need to know its changing character. A good understanding of this book should make you more weather wise than the average meteorologist giving forecasts on the evening news.

Private Pilot Practical Test Standards
U.S. Government Printing Office publication

This FAA booklet states, in terse statements, the flight-test's expectations in terms of performance and evaluation. Not a catchy title, nor is the writing catchy, but it is a "must read" for the flight-test applicant. There are three good reasons for obtaining and studying the publication:

1. There is little here that we have not already discussed in detail, true. But the booklet *is* the official guide to the flight test, stated in the FAA's own words. And if I were you, I would read every word.

2. The booklet provides an excellent review of our discussions together. Here's a suggestion for making the best use of it: As you read each of the booklet's meager statements, take the time to review our in-depth discussions concerning each statement and imagine how you are going to explain and demonstrate the concepts during your flight test.

3. Acquiring the *Private Pilot Practical Test Standards* provides a reward. The FAA has thoughtfully left the back cover blank — just right for the autographs and comments by instructors and fellow student pilots who have traveled the adventure with you. Not a bad souvenir to keep as a reward for a job well done. I only wish that I had retained my own first copy, so adorned.

Again, my very best wishes for your successful flight test and a long, enjoyable career as a certified private pilot.

Index